Statistics for Engineering and Physical Science

Series Editors
P. Green, J.F. Lawless, V. Nair

Springer
New York
Berlin
Heidelberg
Barcelona
Budapest
Hong Kong
London
Milan
Paris
Santa Clara
Singapore
Tokyo

Statistics for Engineering and Physical Science

Hawkins/Olwell: Cumulative Sum Charts and Charting for Quality Improvement.

Douglas M. Hawkins
David H. Olwell

Cumulative Sum Charts and Charting for Quality Improvement

Springer

Douglas M. Hawkins
Department of Applied Sciences
University of Minnesota
St. Paul, MN 55108
USA

David H. Olwell
Department of Mathematical Sciences
U.S. Military Academy
West Point, NY 10996
USA

Library of Congress Cataloging-in-Publication Data
Hawkins, Douglas M.
Cumulative sum charts and charting for quality improvement / Douglas M. Hawkins, David H. Olwell.
p. cm. – (Statistics for engineering and physical science)
Includes index.
ISBN 0-387-98365-1 (hardcover : alk. paper)
1. Quality control – Charts, diagrams, etc. 2. CUSUM technique – Charts, diagrams, etc. I. Olwell, David H. II. Title.
III. Series.
TS156.H36 1997
658.5′62 – dc21 97-37929

Printed on acid-free paper.

Production managed by Anthony Battle; manufacturing supervised by Thomas King.
Photocomposed copy prepared from the authors' LaTeX files.
Printed and bound by Braun-Brumfield, Inc., Ann Arbor, MI.
Printed in the United States of America.

9 8 7 6 5 4 3 2 1

ISBN 0-387-98365-1 Springer-Verlag New York Berlin Heidelberg SPIN 10523759

Preface

Quality improvement is an ever-present concern. Applications are found in the manufacturing environment in which it arose originally, and have spread as well to fields ranging from education to health care to government. Statistical Process Control (SPC) is a collection of tools and of ways of thinking about data. Its most obvious application is to literal "processes," for example, automobile assembly lines. A classical use might be to measure the thickness of the paint on each of a random sample of five automobiles per shift, and put this stream of process readings through SPC analyses. Doing so could indicate whether the paint application process was as it should be, or was suffering from some problems that caused the pain thickness to be different than it should.

SPC is not confined just to manufacturing processes, however.

- Document filing is an essential if unglamorous part of most offices. Some degree of misfiling is unavoidable, but it is important to keep the degree small, and to try continuously to decrease it even further. This problem can be addressed by SPC methods, by daily random samples of documents filed, checking for errors. This stream of data can be analyzed using SPC methods to pinpoint particular problem areas – for example, types of documents that present unusual difficulties, or whether some staff need additional training.
- Health care is another area finding benefits from SPC ideas. Time taken for a blood sample to be turned around can be monitored, for

example, and the data used to identify ways of reducing the turnaround. Key quality attributes of patient outcomes and customer satisfaction provide opportunities to apply SPC, as well.

- Managers are using SPC to monitor misconduct, such as sexual harassment and racial discrimination. This allows them both prompt warning of changes to the status quo and successful defense of existing policies and procedures, depending if the organization is in control or out-of-control.

- Short run processes, historically excluded by many SPC methods because of lack of accurate parameter estimates, are finding new short-run methods applicable.

Some applications of SPC ideas seem to have nothing whatever to do with processes. Consider multiple regression, in which we seek to predict some dependent variable from a set of predictor variables. A major concern is whether the regression model holds–for example, if using linear regression, whether the dependent variable does indeed vary linearly with the predictor variables. Some of the methods that have been proposed for this testing use SPC methodologies–in particular the CUSUMs of this book.

The reach of SPC methods is therefore very broad. We frame much of the discussion in traditional process terms, as if the data were like the paint thickness example. You should not take this as a straitjacket, however: to be applicable, all the CUSUM methodology needs is a suitable statistical model for the data.

Audience and scope

This book is aimed at a range of people: students, practitioners, quality professionals, and statisticians. Individual chapters are layered, with the material important for application first, and any deeper technical discussion later on. To keep the flow of discussion smooth, the main body of text does not have many detailed literature references or attributions; instead, we provide key references and further reading at the end of each block of material.

Chapters 1 through 5 cover the essentials of setting up and running CUSUMs for the most common applications. Those who want to apply CUSUM methods without necessarily going deeply into the theory or methodology should be able to find what they need in these chapters.

Chapter 6 sketches the theoretical underpinnings of CUSUMs and assumes some background in the theory of statistical inference.

Chapter 7 deals with the not-uncommon situation that you either have short runs (and so can not gather large calibration data sets ahead of implementing the charts) or that you prefer to start charting right away, and do not want to wait for the collection of a large volume of calibration data. This is important for both practical and theoretical reasons, and so everyone should read it.

Chapter 8 covers the situation where you have a collection of different but related process measurements (for example, room temperature and humidity) and want to monitor them together, rather than as separate SPC problems. Reading this chapter requires some background in multivariate statistics.

Three more specialized topics in Chapter 9 round out the book.

Software

Modern SPC is usually carried out on computers. A Web site has been set up to help with this. It contains computer programs for calculations needed to calibrate CUSUMs and measure their performance, and also has stand-alone programs to compute CUSUMs from user-supplied data files. The sample data sets discussed in the book are also on the web site.

We also provide spreadsheet files which allow the user to implement many of the techniques in the book easily. A rule of thumb: if we wrote any macros, Fortran code, or data files to produce the illustrations or examples in the text, you can find them at our web site. You are welcome to them, with the standard exclusions of liability and disclaimers which our attorneys insist we attach.

We discuss this software briefly in the appendix.

We have carefully attempted to remove all errors from this monograph. However, should any be discovered, we will post an errata list on the web site. Suspected errors may be brought to the attention of either author.

Thanks

We appreciate the support of our sponsoring editor, John Kimmel, at Springer-Verlag, who has patiently assisted us during the years this monograph was in production. We are similarly appreciative of our wives, Greer and Karen, who have also patiently tolerated our efforts.

Steve Charbonneau of the Department of Systems Engineering, USMA, helped one of the authors learn Visual Basic for Applications, which was used in the Excel spreadsheets provided on the web site for the text.

Three anonymous reviewers made very helpful comments during the editorial process, which resulted in an improved text.

Disclaimers

Software and algorithms

Springer-Verlag or the author(s) make no warrant or representation, either express or implied, with respect to this software or book, including their

US Government

U.S. Army regulations require the second author to include the following statement:

> The views expressed in this book are those of the author and do not represent the official policy or position of the United States Military Academy, the Department of the Army, the Department of Defense, or the United States Government.

Douglas M. Hawkins
St. Paul, Minnesota

David H. Olwell
West Point, New York
October 1997

Contents

List of Figures

1 Introduction

Happy is he who has been able to learn the causes of things.

Virgil

1.1 Common-cause and special-cause variability

Suppose you have some "process" whose ongoing quality you want to assess. You do this by making regular readings on some measurable property of the process. Some examples might be:

- You run a sugar packaging plant where a continuous filler line fills paper bags with sugar. Each bag is supposed to contain 10 pounds of sugar. Although the inevitable random variability makes a constant weight in all bags impossible, you would like to check that there is no excessive variability from one bag to another, and that the average weight of all bags is correct. To achieve this, you take random samples of the production from each shift and weigh the sugar in each sampled bag accurately. These weights are your process measurements.

- You run the emergency room in a hospital. There is some concern about the time taken with the paperwork admitting accident victims. To check this, you have an observer watch a random sample of incoming accident victims and see how long it takes to fill out the forms for each of them. It is clearly important to have a sample that represents all different times of day and different days of the week, so you ensure that your sample correctly represents all these different time periods.

This second example differs from the first in several respects. The most important of these is not the very different context of a hospital rather than a factory; this is a minor issue. The major one is that although the sugar packaging machine may well be expected to run much the same regardless of time of day or day of the week, the same is not true of hospital admissions. An important part of analyzing the emergency room data will consist of finding a good "baseline" model that accommodates the normal variation by time of day and day of the week, as well as normal but erratic features such as holiday weekends.

- As an apparently completely different application, you are looking at the relationship between the unemployment level and real salaries. You use some form of multiple regression for this, using quarterly data. You are concerned that the relationship between these two variables may have changed over time. You can check this using the methods of "recursive residuals" in which you start out with the earliest data, repeatedly add one time period to the database, refit the model, "predict" the next observation, and then monitor the agreement between the actual and predicted values. Despite the total lack of any apparent connection between this problem and a process, statistical process control (SPC) methodologies have been applied very successfully to this problem of model diagnosis.

In all these examples, there is inherent random variability in the process readings. Let's consider the implications of this. The sugar bags of the first example are labeled and sold as containing 10 pounds of sugar. In fact, the amount of sugar in a bag varies from one bag to another. Part of this variability is purely random; it is unpredictable and cannot be reduced without making some process change — for instance, by modifying the equipment used to measure the sugar into the bag. Some might be more systematic and predictable. The weight of sugar in a bag might be affected by the humidity at the time of packing. This type of variability could be dealt with in a number of ways. The humidity could be controlled more closely, or the machine's settings could be tuned in real-time to counteract the effect of changes in humidity.

Another possibility is to do nothing about controlling or correcting for humidity, but to decide to accept the higher level of variability in the fill weight of the bags. The variability in bag weight creates obvious difficulties for the manufacturer. To label bags as containing more sugar than they do is illegal and unwise. You can sidestep this problem by deliberately setting your packaging machinery so that the *average* bag contains more than 10 pounds of sugar, thereby creating some leeway before the random

Applications of SPC:

- Manufacturing
- Service industries
- Management
- Model diagnostics

Variability in processes:

- Inherent
- Source of quality problems
- Attributable to *special* and *common* causes

variability will give unacceptable numbers of underweight bags. Clearly the greater the variability among bags, the larger this cushion of additional free sugar will need to be to have only a few underweight bags. The variability in the weight translates directly into additional free sugar given to the consumer. There is an immediate (and in this case easily quantified) benefit to the producer in reducing the bag to bag weight variability.

More generally, uncontrolled random variability is a major source of quality problems, although in many applications its cost is not easy to quantify. Consider, for example, the 750-hour electric light bulb. Its actual life varies randomly and some light bulbs will last well above this nominal level, whereas others burn out more quickly. It is the light bulb that burns out prematurely that the consumer remembers, not the one that is still working at 2,000 hours. The light bulb manufacturer also has an incentive, although not one easily measured in dollars and cents, to make a product with more consistent predictable performance.

Looking into this a little more deeply, reducing variability in the life of light bulbs is likely to lead to an increase in the average life also (since the life of the bulb is strongly affected by sections where the filament is thin), so reduction in variability in this case may lead directly to a better average product.

Another, more traditional, example is the turning of engine bolts for automobiles. If a bolt has a nominal diameter of 5mm, then its fitness for use relates directly to variability from this target. Bolts that are too broad or too narrow are unfit. No two bolts will have exactly the same diameter; there is variation from one bolt to another in the production process.

This bolt-to-bolt variability can be broken down into various sources. If the bolts are made by two different machines and these machines are set up differently, then the bolts from one machine may be systematically bigger than those of the other. As the cutter making the bolts wears, the diameters may change systematically. If the feedstock changes to a harder steel, this change may shift the diameters too. These changes are examples of *special* causes. They all cause heterogeneity in the bolt population. This heterogeneity could be accounted for, or even undone, by separating out the streams of bolts made by the different machines, or those using the different feedstocks, or those made shortly after cutter adjustment from those made later. Finally, superimposed on these identifiable causes of variability in bolt diameter, there will be purely random variability from one bolt to another. This random variability is ascribed to a catch-all category of *common* causes.

We concentrate for the moment on the situation where only common cause variability is present. If we take a sample of bolt diameters, then they will look like a random sample from some statistical distribution. The variability could be summarized by some central value (for example, the mean) and the spread (for example, standard deviation) of the distribution of the bolts. Both the mean and the standard deviation are important in producing bolts of the proper diameter. If their true mean diameter is different than the nominal 5 mm, then the bolts will tend systematically to be too small or too large, and either way they will not be as good as they could be. The diameters would also be unacceptable if the standard deviation was too large, for this would lead to excessive numbers of bolts "out in the tails."

Both elements (mean and spread) are important for quality. We could reduce variability and still have a poor product because of mis-centering. The two problems can be uncoupled in this problem (and many others) because there are machine adjustments that can be made to re-center the process. Making engine bolts of the right diameter involves two separate steps: controlling the variability from bolt to bolt, and keeping the center of the distribution (the true mean) constant and at the level that will minimize problems of off-specification product.

We have introduced the concept of *special cause* and *common cause* variability. A widely held tenet of statistical process control is that proper monitoring of the process and follow-up action can potentially identify and remove special causes of variability, but that reducing the common cause variability requires a fundamental change in the way the process operates. Chipped cutter tools, changes in the hardness of the feedstock, and errors in calibration are all examples of special causes that could cause the mean diameter of manufactured bolts to change. Detecting them and

Removing special causes **improves process quality**.

removing their effects will keep the mean diameter more nearly constant and so reduce overall variability. Common cause variability is what is left when these special causes have been accounted for and removed. Reducing this variability would require more fundamental changes — for example, replacing lathe bearings with others made to tighter tolerances, or finding a new supply of feedstock with less variable physical properties. Nevertheless, the goals of continuous quality improvement mean that, once special cause variability has been tracked down and removed as far as possible, further efforts will have to be made to reduce common cause variability.

1.2 Transient and persistent special causes

SPC methodologies concentrate on providing tools to help in detecting and diagnosing special cause variability. It is helpful to recognize two distinct types of special cause variability. One is *transient* in its effect. These special causes may affect the process for a short while, then disappear only to reappear at some future time. Examples might be:

- a process using external electrical power. If there is a thunderstorm in the area, electrical disturbances in the line voltage could affect the process and the resulting measurements. This special cause would last for the duration of the storm and then go away, recurring at the next storm. A close relative of this problem is the factory where switching one machine on or off creates transients in the whole factory's electrical system, possibly affecting other processes.
- Automatic teller machines, assessed by their transaction times. Most ATM users are fluent in their use of the machines and so their transaction times fit to a common statistical distribution. A small fraction of the transactions, however, are made by people using an ATM for the first time, and their transaction times are likely to be far longer.
- As an apparently completely different problem, we might be interested in exploring the relationship between interest rates and unemployment using regression on quarterly data. In a quarter following a major earthquake, this relationship would be disrupted. In time, though, the relationship might return to its longer-term form.

The other type of special cause is one whose effects persist until the problem is detected and diagnosed. Examples might be:

> Transient and persistent special causes are best diagnosed by different methods.

- if we chip the cutter tool making our 5mm engine bolts, then the diameters of the bolts may undergo a step change, and will remain at this changed level until the problem is discovered and cured.
- A new operator may control equipment differently than the previous operator. This will lead to a shift in the process. The shift may persist until the difference between the operators is discovered and some action taken to remove it.
- Many chemical assay methods are comparative, providing a measurement that is relative to some standard "reference" material, which is assumed constant. When the reference material is used up, it will have to be replaced by a new supply. Any errors in relating the true value of the new reference material to that of the old will translate into a systematic difference in the assays produced by the method.

There are, of course, other causes intermediate between these extremes: causes whose effect persists for a time but then will change or go away, even if not treated. It is nevertheless helpful to keep this broad distinction between transient and persistent special causes.

Both persistent and transient special causes are important, and we need good methodologies for detecting both if we are going to remove them and improve the process. Since the two types of causes leave different footprints in the process measures, we need different methodologies if we are to detect and diagnose both types effectively.

1.3 The Shewhart and CUSUM charts

The best-known control charts are those pioneered by Walter Shewhart (1931), and are exemplified by his Xbar and R charts. Consider again a process making 5mm bolts. There will be variability from one bolt to another. Suppose studying this variability has shown that the diameters of bolts appear to follow a normal distribution with true mean $\mu = 5$mm and standard deviation $\sigma = 0.1$mm. Then a common Shewhart control chart for the mean might be constructed as follows.

- Take a random sample of size 5 bolts from each shift, and measure the diameter of each.

- Compute $\overline{X}$ — the average of the 5 diameters, and R — the range from the smallest of the diameters to the largest.
- Plot $\overline{X}$ on the Xbar chart.
- Plot R on the R chart.

The Xbar chart is just a time-ordered plot of the $\overline{X}$ values from different shifts, with three drawn-in horizontal lines. There is a *center line* at the true in-control mean of 5mm, and two *control limits*: an *upper control limit* and a *lower control limit*. These control limits are located 3 standard errors of the mean above (upper) and below (lower) the center line. As the standard error of the mean of a sample of size n is $\sigma/\sqrt{n}$, these limits for our bolt example are set at $5 \pm 3 \times 0.1/\sqrt{5}$, or 5 ± 0.13.

If the latest $\overline{X}$ is above the upper control limit, then the Xbar chart is considered to have given the signal that the process has gone *out of control*, that is, that the data are no longer plausibly a random sample from the *in-control* normal distribution with mean 5 and standard deviation 0.1 that they should be. An investigation into the cause can then be made.

The Xbar chart monitors possible changes in the mean. To complete the picture, we also need some way to monitor possible changes in the spread of the diameters. This is done with the *R chart* , a time-ordered plot of the sample ranges. Like the Xbar chart, this has a center line at the expected value of the range of a sample of size 5 from a normal distribution with standard deviation of 0.1 — this center line is at 0.23. There is also an upper control limit (in this case at 0.49) but for this sample size there is no lower control limit. If the latest R value plots above the upper control limit, then the R chart is considered to have given the signal that the process has gone out of control.

Figure 1.1 gives an example of the Shewhart Xbar chart for this situation. It shows simulated data on 100 bolts, 5 taken in each of 20 rational groups.

Note that all points remained well inside the control limits, leading to the conclusion that the process was in control, and that it should therefore be left to run unaltered. (The Shewhart R chart for this data set is not shown here, but its points also lay well within its control limits.)

The Shewhart chart has a beautiful simplicity to it. It may also be as valuable for what it prevents as for what it motivates. As long as the points plot inside the control limits, no action is taken to alter the process. This rule can stop much unproductive tinkering that could take a process from a good state into a bad one. The control limits are placed sufficiently far from the center line that very few samples should plot outside them if the process remains at its in-control distribution. This means that when the control chart does give a signal, it should be taken seriously.

These attractions of the Shewhart chart should not blind one to a serious limitation. It has no memory, and so although it is very effective for detecting isolated special causes that lead to large shifts in the data, it

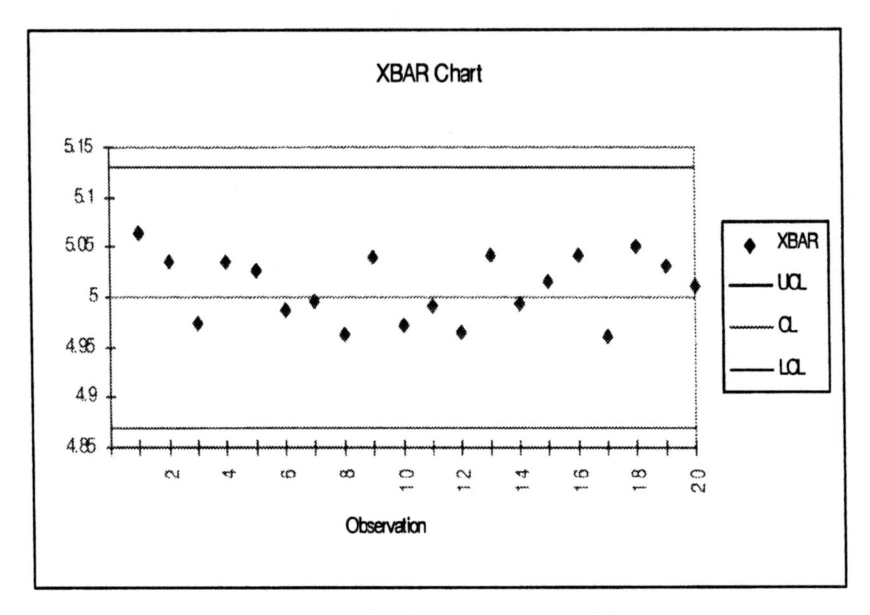

FIGURE 1.1. Shewhart Xbar chart

Shewhart Control charts:

- Detect transient special causes
- Prevent overreaction to usual process variation
- Are memoryless

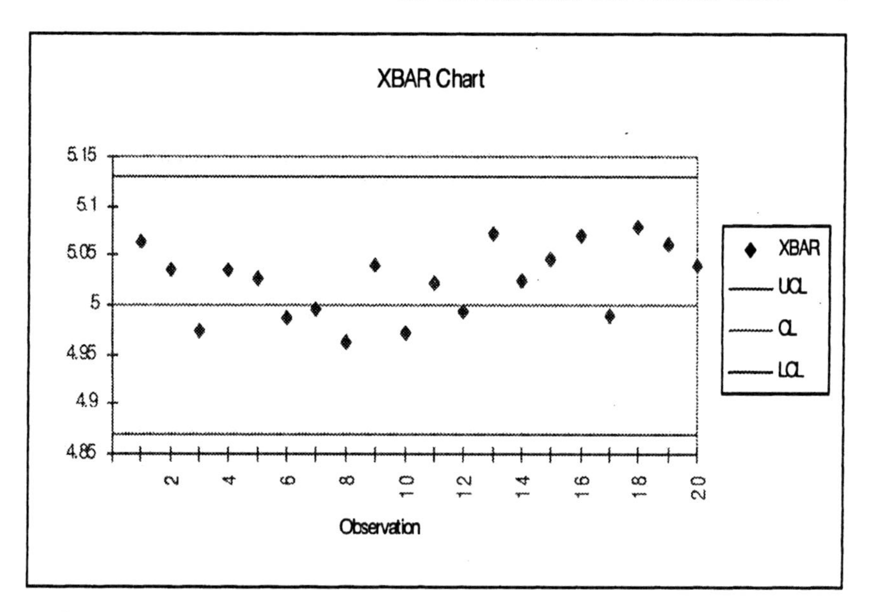

FIGURE 1.2. Shewhart chart of diameters, altered from Figure 1.1. The last ten readings have been increased by 0.03 mm. The chart does not detect this small persistent shift.

is not very effective in detecting more moderate shifts, *even if* these more moderate shifts persist.

There have been numerous attempts to patch this deficiency in the Shewhart chart by adding "supplementary runs rules" — rules that will lead to a signal if the successive points have specified unusual patterns to them, even if they all remain inside the control limits. One example is the venerable rule that signals that the process is out of control if two out of three successive points plot more than two standard errors from the center line.

These attempts are generally less successful than going back to first principles and finding a different technology that is inherently better suited to the detection of persistent shifts in the stream of process data. As such a technology would need to accumulate information from successive readings, the idea of cumulative sum (CUSUM) charting was recognized early on as intuitively attractive for detecting smaller but persistent shifts.

As an illustration of the problem and a possible solution, look at Figure 1.2. This shows the same sequence of bolt diameters we saw in Figure 1.1, except that the last 50 diameters (and so the last 10 Xbar values they give rise to) have all been increased by 0.03mm. All the points on the Xbar chart remain well within the control limits, and there is little compelling evidence of a shift in the data.

There is another diagnostic that is much more effective for detecting persistent small shifts in mean. For this diagnostic, we string the data

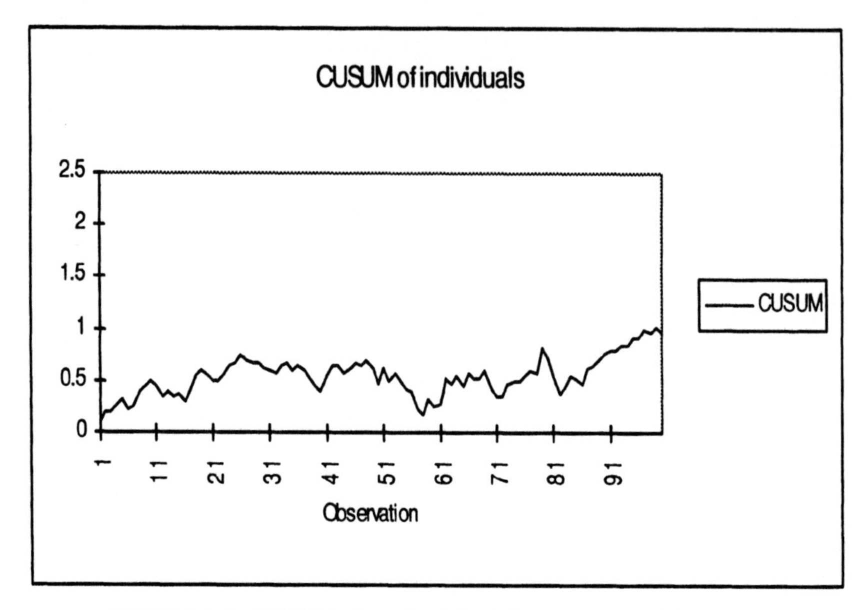

FIGURE 1.3. CUSUM plot of original diameters from Figure 1.1.

out into a series of 100 diameter readings, rather than thinking of them as comprising 20 "rational groups" of size 5. We motivate this change of viewpoint later. Write X_j for the jth of these readings, and define the cumulative sum (CUSUM):

$$C_n = \sum_{j=1}^{n} (X_j - \mu) \tag{1.1}$$

This is a running total of the deviations of the first n readings from the in-control mean μ — in this case 5mm. The CUSUM consists of a plot of C_n against n.

Figure 1.3 shows the CUSUM plot of the original 100 bolt diameters charted in Figure 1.1, and Figure 1.4 is the CUSUM of the shifted bolt diameters of Figure 1.2.

The two CUSUMs are on the same scale, making it easier to compare them. Although the differences between the Shewhart Xbar charts of Figures 1.1 and 1.2 were hardly compelling, the CUSUM charts of Figure 1.3 and Figure 1.4 are strikingly different. The CUSUM of the original unshifted data series is a largely featureless plot drifting across the page. The CUSUM of the shifted diameters starts out drifting across the page, but then changes and moves strongly upward.

This transition from a trendless drift to a near-linear drift is the characteristic feature of the CUSUM of a process whose center has shifted. Later we discuss formal procedures for distinguishing genuine drifts from back-

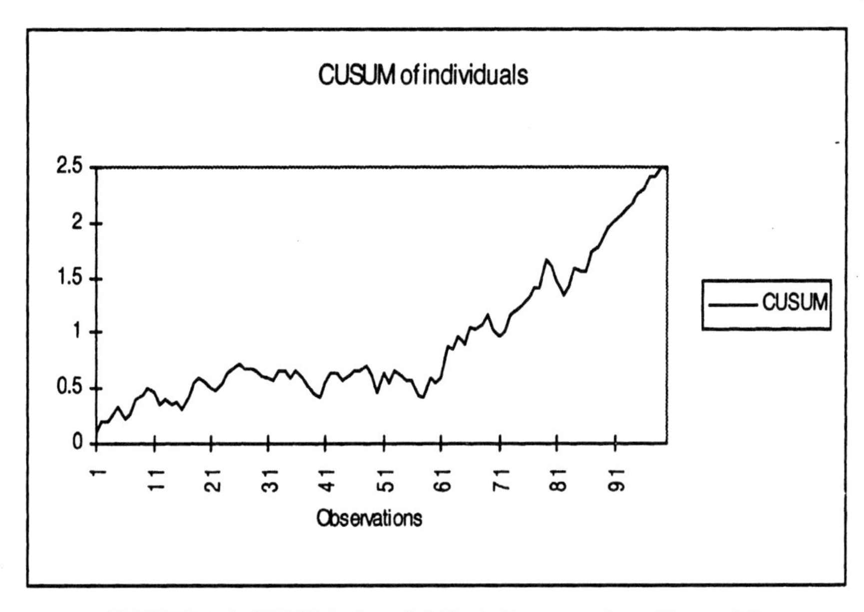

FIGURE 1.4. CUSUM plot of shifted diameters from Figure 1.2.

ground random variability, but for the present it is enough to note that the change seems visually clear, and that it seems to have occurred a bit before the sixtieth observation in the series.

1.4 Basis for the CUSUM chart for a normal mean

Let's look in a bit more detail at the statistical underpinning of the CUSUM chart. Suppose that while the process is in control the readings X_n are statistically independent, and follow a normal distribution with known mean μ and known standard deviation σ. This statistical model has three parts.

- The readings are statistically independent and identically distributed.
- The readings follow a normal distribution.
- The true mean and standard deviation of this normal distribution are known exactly.

None of these three assumptions can be taken for granted. It is not uncommon for readings taken close together in time to have some degree of association even in assembly-line manufacturing, and serial association is the usual state of affairs for continuous-flow chemical processes. We do not spend a lot of time on this issue right now except to remind you that it is important to check the lack of correlation. Later we discuss some techniques for making the necessary checks, and for CUSUMming readings from processes with large "random batch effects."

The second assumption is noteworthy because we do not usually see a parallel requirement in discussion of the classical Shewhart chart. This is not because the Shewhart chart is immune to the effects of the distribution of the data (it isn't) but because knowing the distribution of the data is essential to figuring out the chart's false-alarm rate and how sensitive it is to actual shifts, and CUSUM workers have tended to pay more attention to these performance issues than have Shewhart users.

The normal distribution part of the assumption can be checked using histograms of samples of previous readings made while the process was believed to be in a state of statistical control. A more powerful check is given by a normal probability plot; many statistical packages will produce this diagnostic.

Although these early chapters concentrate on the CUSUM for monitoring the mean of normal data, the CUSUM is by no means tied to the normal distribution, and we introduce CUSUMs for many other nonnormal distributions later.

Finally, there is the assumption that you know the exact values of the parameters μ and σ. This is never the case, but it is possible to get very precise estimates. In practice, the parameters are estimated by measuring the process for a suitably long period while it is in a state of statistical control, and using this data set to calculate estimates of the parameters. These estimates are then taken to be the true parameter values. Provided the calibration data set was large, these estimates will be sufficiently close to the true parameter values for most practical purposes. In Chapter 7 we discuss "self-starting" CUSUMs in which we escape this assumption of parameters being known exactly prior to the start of control charting, but the traditional assumption of known parameters is fine for now and allows us to develop the ideas.

Before leaving this topic, we should warn that the samples needed to calibrate CUSUM charts may be much larger than those needed to calibrate Shewhart charts. The reason for this is that Shewhart charts are insensitive to small shifts in the parameters, and so they are not much affected by modest errors in the specified parameter values. CUSUMs, by contrast, are designed to detect persistent but possibly small differences between a process's current parameter values and the values specified for the in-control process. Since a slightly misspecified value for the in-control parameter will mean that there is a persistent difference between the specified value and the process's true current parameter values even when it is in control, any random error in the estimated parameter values will tend to cause false alarms. Calibration samples on the order of a few dozen process readings may be enough to set up a Shewhart control scheme, but are often far too small for calibrating CUSUMs, which need to be calibrated using sample sizes in the hundred scale rather than the dozen scale. We discuss this in Chapter 7.

The CUSUM C_n can be defined in two functionally equivalent ways. One, in the original scale of the problem, is the definition already used:

$$C_n = \sum_{j=1}^{n}(X_j - \mu)$$

The other involves first *standardizing* the readings to have zero mean and unit standard deviation. We define

$$\begin{aligned} U_j &= (X_j - \mu)/\sigma \\ S_n &= \sum_{j=1}^{n} U_j \end{aligned}$$

It is clear from inspection that the CUSUM C_n is the CUSUM S_n scaled by a factor of σ, the standard deviation of the readings. Thus the CUSUMs of C_n and of S_n are identical except for the units of the vertical axis. The vertical axis of the S_n CUSUM will be measured in multiples of the standard deviation σ of the data, whereas the vertical axis of the C_n CUSUM will be measured in the same units as X. Statistically, the two CUSUMs contain the same information.

Opinions differ on whether you should use standardized or unstandardized CUSUMs. The advantage of standardizing is that the units of the vertical axis are absolute. This means that the standardized CUSUM will be the same whether X is measured in meters, millimeters, inches, or light years. It also allows you to compare CUSUMs of entirely different quantities; for example, using standardized data you could plot a CUSUM of room temperature and one of humidity on the same scale and compare their detailed shapes.

The advantage of not standardizing is that the units of the vertical axis are units of the original measurement, and so may be more easily interpreted.

Neither scale has advantages so great as to justify any claims that it is clearly superior to the other; so deciding between standardized and unstandardized data for CUSUMming can be a matter of individual preference, and can lead to different choices for different problems. We are not rigid on this issue, but use standardized or unstandardized data — whichever seems more sensible for the particular problem at hand.

1.4.1 Statistical properties of the CUSUM

Turning to the statistics of the CUSUM, C_n is the sum of independent normal $N(0, \sigma^2)$ quantities. Its distribution is therefore

$$C_n \sim N(0, n\sigma^2), \tag{1.2}$$

The standard deviation of C_n increases with n, being proportional to the square root of n. This means that as n increases, C_n is increasingly likely to be far from zero. This has implications for the mechanics of plotting a C_n CUSUM. Even while the process is in control, as n increases the CUSUM requires an ever wider expanse of paper (or of computer screen if the CUSUM is being displayed rather then drawn physically).

The equation for C_n shows that it can be written in a recursive form:

$$\begin{aligned} C_0 &= 0 \\ C_n &= C_{n-1} + (X_n - \mu). \end{aligned}$$

This form shows the easiest way to compute the CUSUM, whether by hand or on a computer. Each point of the CUSUM is the previous point plus the offset of the latest point from μ. The recursive form is also useful for theoretical studies of the CUSUM, since it shows that the process C_n is a random walk.

The corresponding recursion for the standardized form of the CUSUM is

$$\begin{aligned} S_0 &= 0 \\ S_n &= S_{n-1} + U_n. \end{aligned}$$

Both forms imply that, while the process is in control, the CUSUM will be a random walk with no drift — each point is the preceding point plus an offset with zero mean. The CUSUM will center on the horizontal axis, and be subject to ever-widening excursions from the axis.

Summary:

- CUSUMs detect persistent changes.
- A CUSUM requires an explicit and precise statistical model for the observations.
- To get a precise model requires extensive historical data.

1.5 Out-of-control distribution of the CUSUM

Suppose that at some instant m the distribution of the X_n changes from $N(\mu, \sigma^2)$ to $N(\mu+\delta, \sigma^2)$; in other words, from instant m onwards the mean

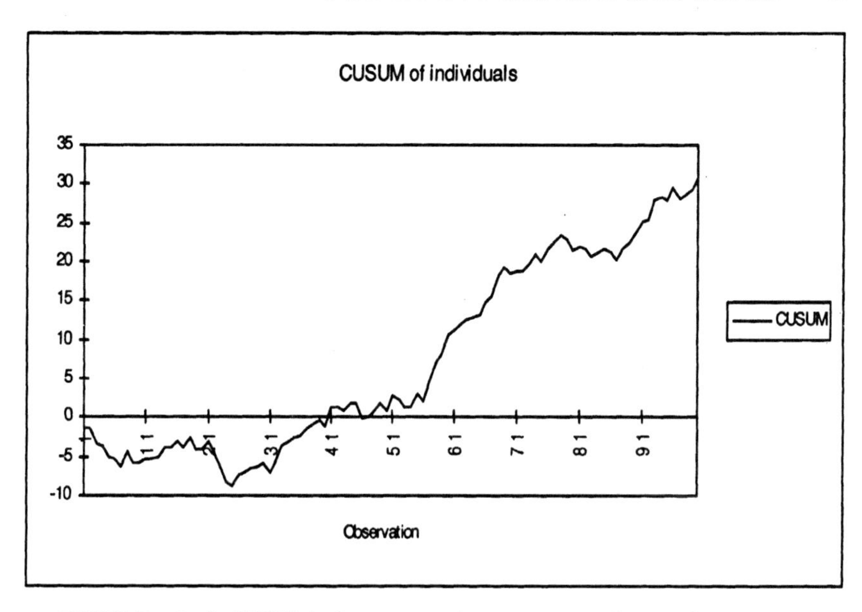

FIGURE 1.5. A CUSUM of a process that went out of control at $m = 50$.

of X_n undergoes a persistent shift of size δ. At any subsequent instant, say n, we can write the CUSUM as

$$\begin{aligned} C_n &= \sum_{i=1}^{n}(X_i - \mu) \\ &= \sum_{i=1}^{m}(X_i - \mu) + \sum_{i=m+1}^{n}(X_i - \mu). \end{aligned}$$

In the second of these two summations, the summand is distributed as $N(\delta, \sigma^2)$. Thus the second sum will have distribution

$$\sum_{i=m+1}^{n}(X_i - \mu) \sim N[(n-m)\delta, (n-m)\sigma^2]. \tag{1.3}$$

In other words, the average value of the CUSUM at time $n > m$ is $(n - m)\delta$. This means that starting from the point (m, C_m), the CUSUM on average will trace out a path centered on a line of slope δ.

Figure 1.5 illustrates a CUSUM of a process that has gone out of control at $m = 50$. The in-control distribution was $N(0, 1)$. The out-of-control distribution is $N(.5, 1)$, so $\delta = 0.5$. Notice the change to linear drift from no drift.

This is the basis for using the CUSUM to diagnose shifts in mean. While the process remains in control and the readings X_n follow the in-control

$N(\mu, \sigma^2)$ distribution, the CUSUM follows a distribution centered on the horizontal axis. If the mean undergoes a step change, then the CUSUM develops a linear drift, and its distribution will center instead on a line whose slope δ equals the shift in mean. The diagnosis of the CUSUM therefore consists of distinguishing the no-drift in-control behavior from the linear-drift behavior following a mean shift.

In principle, the mean could change in more complex ways and the CUSUM could diagnose these more complex shifts. Suppose that instead of a step change the mean develops a slow linear drift ; specifically, suppose the mean of X_n for $n > m$ was $\mu + (n-m)\delta$. Then the behavior of the CUSUM beyond the change point would be:

$$\begin{aligned} C_n &= C_m + \sum_{i=m+1}^{n} (X_i - \mu) \\ \sum_{i=m+1}^{n} (X_i - \mu) &\sim N\left(\frac{(n-m)(n-m+1)\delta}{2}, (n-m)\sigma^2\right). \end{aligned}$$

The linear drift in the mean of X translates into a quadratic drift for the CUSUM.

In practice, it is not easy to distinguish the (locally approximately linear) quadratic drift of this slowly changing mean from the linear drift caused by a step change, so this possibility is of theoretical rather than practical value. Although the CUSUM will respond to the slow drift in mean by signaling, it is not easy to differentiate it from a step change in mean unless you are prepared to wait until quite long after the change.

We return to the example in Figure 1.5, except we now allow for linear drift. In other words,

$$X_i \sim N((i-50).05, 1), i > 50.$$

We plot the resulting CUSUM in Figure 1.6. Although the graph is really parabolic, it is not easy to distinguish this curvature from the linear behavior of Figure 1.5 until after about $n = 90$.

1.6 Testing for a shift — the V-mask

The behavior of the CUSUM in and out of control shows us what sort of behavior would constitute evidence of a shift in mean: a change from generally horizontal motion to a non horizontal linear drift. The historical formal tool for making this determination is the V-mask, so named because of its shape. We motivate this initially as being intuitively reasonable, and later show that it has more than just an intuitive basis.

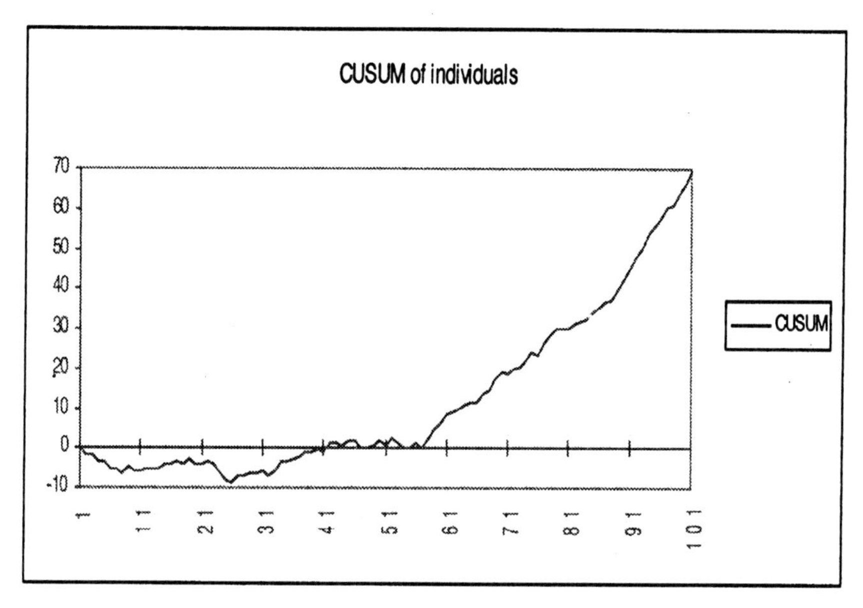

FIGURE 1.6. A CUSUM of a process undergoing slow linear drift in the mean after observation 50.

It simplifies things slightly to work with the standardized form of the CUSUM S_n, and since C_n is just S_n multiplied by σ, nothing essential is lost by doing this.

Suppose we are at point m and want to know whether it is the start of a non-horizontal linear drift. Fix upon a particular slope for the non-horizontal drift we are looking for; to fix upon some value, let's suppose we fix upon the drift of slope 0.8. This corresponds to a shift in mean of 0.8 standard deviations. So we want to decide between a slope of zero, and a slope of 0.8 for the segment leading out of the point $\{m, S_m\}$. A reasonable way to do this is to draw a line of slope midway between these two levels, that is of slope 0.4. If the CUSUM remains close to or below this line, you will conclude that the mean has not shifted. But if the CUSUM goes substantially above this line, that would be indicative that the drift of slope 0.8 is more believable than is slope zero.

We make this concrete by saying that if any future point is more than some threshold height h above this line, that will constitute the evidence of the shift.

This leads to a line of slope 0.4 located h units vertically above the point $\{m, S_m\}$. If any future point $\{n, S_n\}$ lies above this line, then that point will signal a shift.

Figure 1.7 shows the CUSUM plot of the shifted data of Figure 1.4, with the reference line above case 50 drawn in. This line could be used to check whether any future case indicated a mean shift that occurred at case 50.

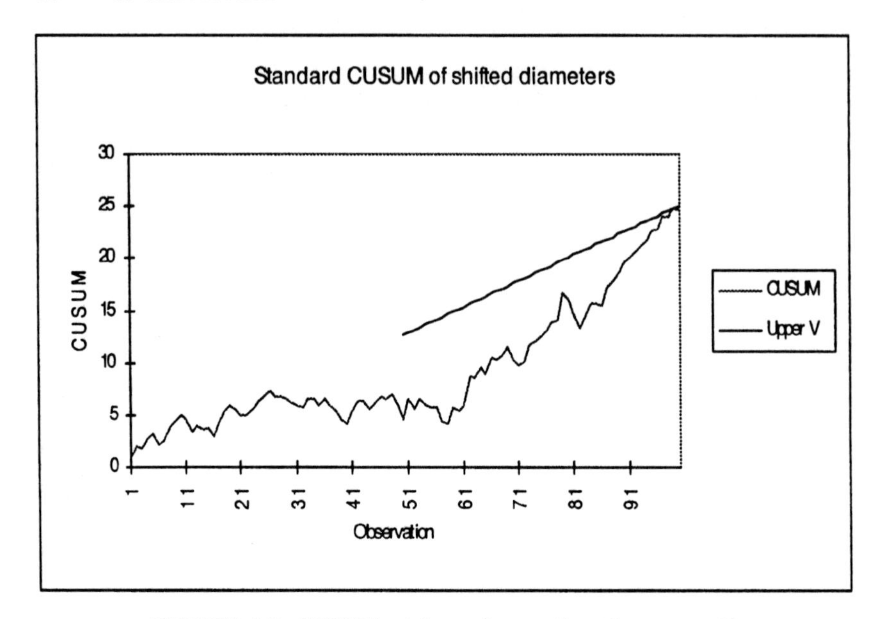

FIGURE 1.7. CUSUM with a reference line above case 50.

The whole series of points happens to lie below this line (although it is drawing very close at the end). This suggests that as of the last case in the data set, this reference line did not yet provide confirmation of a shift starting at case 50. Of course this may just indicate that the shift appeared to start, not at case 50, but at some other case.

In drawing Figure 1.7, we focused on case 50 and the question of whether there was an indication of there being a shift at some future point. Changing focus from the left end of the segment to the right end, it is exactly equivalent to say that, from each future point $\{n, S_n\}$ we draw a line of slope 0.4 starting h units below $\{n, S_n\}$ and going backward. If any preceding point lies below this line, then we conclude that there has been a mean shift.

Putting this in more general symbolic terms, suppose we are interested in seeing whether there has been a mean shift of size Δ standard deviations in the U_j. Draw a line of slope $k = \frac{1}{2}\Delta$, starting from $\{n, S_n - h\}$ and going backward. If any preceding point lies below this line, then we signal an upward shift in mean.

Now, if we are concerned about the possibility of an upward shift of size Δ, then we are presumably also concerned about a possible downward shift of size Δ. We could check for this shift in exactly the same way, with a line of slope $-k$, starting from $\{n, S_n + h\}$ and going backward. If any preceding point lies above this line, then we would conclude that the mean had shifted downward.

A V-mask is designed to best test for shifts of a given magnitude Δ.

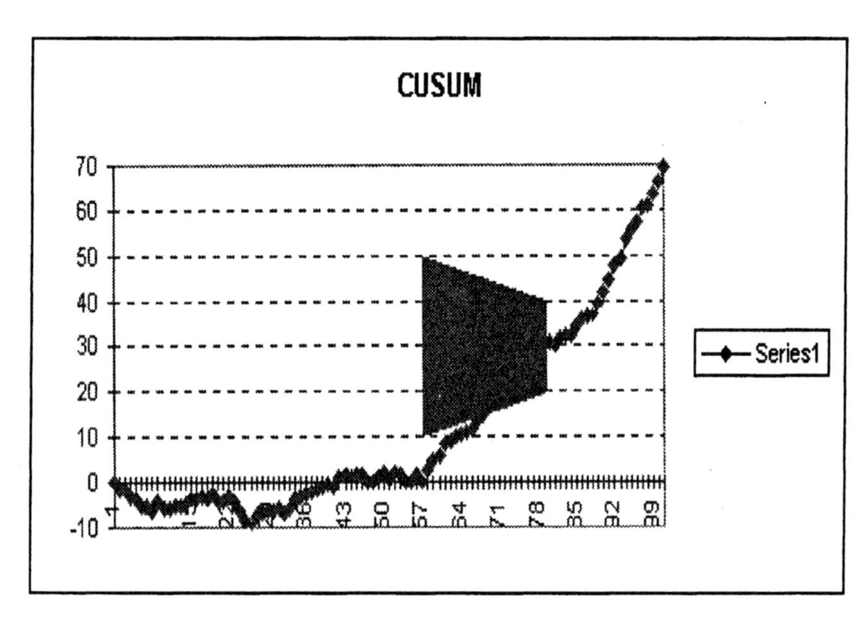

FIGURE 1.8. A V-mask applied to the CUSUM of Figure 1.6 at observation 80. The V-mask signals that the process has been out-of-control since observation 70. Earlier applications of the V-mask also would have signaled out–of–control.

The check on shifts in means then consists of these two lines: one of slope k and passing through $\{n, S_n - h\}$, and the other of slope $-k$ and passing through $\{n, S_n + h\}$.

Finally, the test for a previous shift needs to be applied, not just once, but every time a new point is added to the CUSUM. It is not desirable to have all these lines drawn in, two fresh lines for each new point in the CUSUM. Instead we can do the same test using a cutout "mask." This mask has a flat front (right) end of height $2h$, and two legs of slope k and $-k$. As each new point is added to the CUSUM, align the center of the front edge with the point just plotted, and see whether all previous points are contained within the mask. If they are, then you conclude that the process is still in control - that its mean has not shifted. If any preceding point projects outside the mask, then you conclude that the mean has shifted. This is illustrated in Figure 1.8.

The mask looks like the letter V lying on its side, with the point cut off. This gives rise to its name, the "V-mask". When constructing a cutout

V-mask to use with a CUSUM plotted on paper, it is a matter of taste whether to make the outside of the mask solid and leave the V-shape as a hole, or to make the V-shape solid. The first allows you to look at the previous in-control portion of the CUSUM (the out-of-control part being invisible as it is under the solid part of the mask). The second choice allows you only to see points that correspond to out-of-control signals. Neither of these two choices of what to see and what to have hidden is clearly better than the other.

As mentioned earlier, some users prefer CUSUM S_n of standardized readings, while others prefer the CUSUM C_n of deviations from the mean on the original scale. The two CUSUMs are essentially equivalent, differing only in their vertical scale, where the C_n CUSUM values are σ times those of the S_n CUSUM. The scale carries over to the mask as well. When working with standardized data, k and h are "dimensionless." If working with the unstandardized CUSUM, both k and h need to be scaled up by the factor σ.

1.7 Estimation following a signal

If all previous points in the CUSUM lie inside the mask, then no signal is given. If, however, one of the earlier points lies outside the mask, then a shift in mean is signaled. At this point, it becomes possible to do some estimation – notably of when the shift occurred, and what its magnitude is. These estimates come out of the fact that, following the shift in mean, the CUSUM starts to center on a line whose slope equals the magnitude of the mean shift.

We estimate m, the reading following which the mean shift occurred, as the observation number of the case that falls outside the V-mask. If more than one of the previous points falls outside the V-mask, then we pick as the estimator of m the point lying farthest below the V-mask.

The magnitude of the shift has an intuitive estimate. When the mean shifts from μ to $\mu + \delta$, the CUSUM acquires a linear drift of slope δ. We estimate δ as the slope of the CUSUM over the out-of-control segment from time m, the instant before the shift occurred, to the time n at which the shift is detected.

Some CUSUM users with training in linear regression assume (wrongly) that this slope should be estimated with some sophisticated approach such as linear regression; this is not the case. The optimal estimator is the slope of the line connecting the first and last points on this segment of the CUSUM, and ignores the position of any intermediate points. The CUSUM slope estimator, in other words, is

$$\hat{\delta} = \frac{C_n - C_m}{n - m}. \tag{1.4}$$

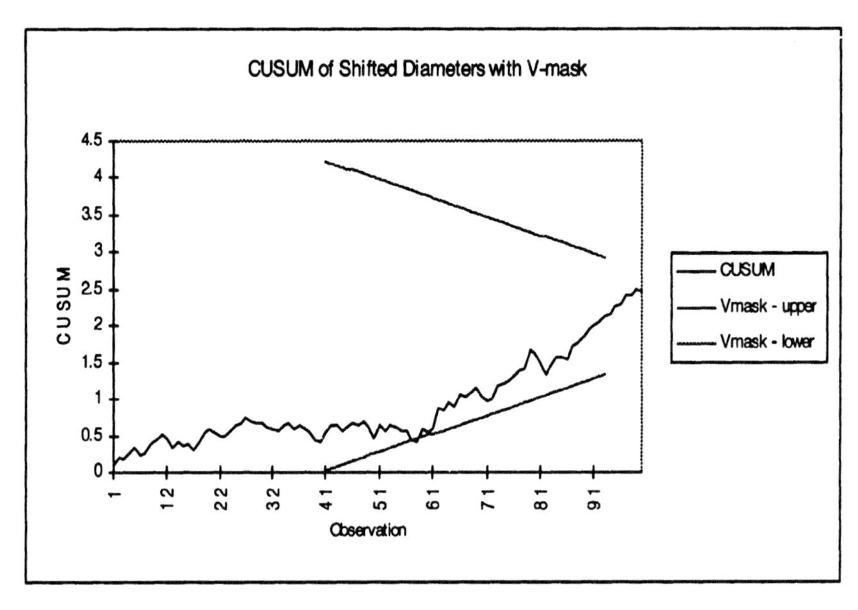

FIGURE 1.9. CUSUM of shifted diameters with V-mask.

Figure 1.9 shows the CUSUM of our example of the shifted series. Superimposed is a V-mask with slope 0.025 (one quarter of a standard deviation) and leg 0.8. The mask is positioned at the first point where the CUSUM gave a signal: this is case 93. At that stage, the value of C_n was 2.13. The point that lies farthest below the mask is case 58, where the value of C_n is 0.42 (these values were not read from the plot, but from the actual calculated values used to produce the plot).

We estimate then that the last in-control point before the shift was case 58, and that the magnitude of the shift in mean is

$$\hat{\delta} = \frac{2.13 - 0.42}{93 - 58} = 0.049.$$

In this case, the data are artificial, so we know where the mean shifted and by how much: it shifted at case 50 by an amount 0.03. The estimate of the time of shift m is 8 sample values away from the correct value. This is not at all bad considering how small the shift is (three tenths of a standard deviation). The estimate of the shift, 0.049, is quite a bit bigger than the actual shift. This is a characteristic of the CUSUM estimate — it is biased away from zero. If the shift is positive, then the estimate from the CUSUM averages a larger positive value than the actual value; if the shift is negative then the CUSUM estimate averages a larger negative value.

This bias in the estimate of the size of the shift means that we need to use it with some caution, expecting that the actual shift will not be quite as large as the CUSUM suggests. Still, even a somewhat biased estimate is better than no estimate at all.

> Summary: A CUSUM signal provides an estimate of both
>
> - when the shift occurred and
> - the magnitude of the shift

1.8 Using individual readings or rational groups

Shewhart charts can be made of individual observations, but seldom are. Usually the chart is of the mean of a "rational group" of observations. In setting up a Shewhart Xbar and R chart for the bolt diameters, standard practice would typically be quite like that used in the example, to take a sample of 5 bolts from each shift and plot the means of the 5 bolts.

When monitoring a process using Shewhart charts, breaking the readings down into rational groups is an essential ingredient of success. The concept underlying this grouping is that it should be chosen in such a way that special causes are likely to affect all or none of the readings in any particular rational group.

Under this assumption, it is reasonable to reduce the readings of a rational group to their average (or some other sensible measure of location) and their standard deviation (or some other sensible measure of scale). Checking the rational group means then indicates whether a particular rational group may have suffered from some special cause that affected the mean of the process at that time. Similarly checking the standard deviation diagnoses whether the process may have been suffering from some special cause that affected the variability of the process at that time.

Rational groups therefore need to be chosen in a way that reflects the likely "reach" of special causes. Doing this effectively is likely to depend on some knowledge of the process.

For example, in a typical three-shift manufacturing operation, a new crew of operators comes on every shift. Since some special causes are likely to be related to the operators, it generally makes sense to define rational groups that are contained within shifts. Thus one might take a sample of four readings during a shift, and reduce the four readings to their mean and standard deviation for charting.

Of course, this is not a good way to detect special causes associated with shift startup. Some processes have a tendency to operate differently at the start of a shift than at the end. If you want to be able to detect this tendency, then the rational grouping by shifts (which implicitly treats all the readings within the shift as exchangeable) will not provide good diagnostics.

CUSUMs have a different motivation. They are intended for persistent shifts that will affect all process readings, and so the idea of rational grouping plays no useful part in CUSUM design and use.

The readings used in a CUSUM chart can be individual readings, or the means of rational groups. As we now illustrate, it does not make much difference to the CUSUM chart whether individual readings or means of rational groups are used.

If individual readings X_i are normally distributed with mean μ and variance σ^2, then the means of random samples of size n, $\overline{X}_i$, say, will follow a normal distribution with mean μ and variance σ^2/n. You can then subject these rational group means to CUSUM control in exactly the same way you would handle individual observations.

If we work with $\overline{X}_n$, the successive means of the rational groups, the CUSUM of the means is given by

$$C_n = \sum_{j=1}^{n}(\overline{X}_j - \mu).$$

The relationship between this CUSUM and one of the individual readings making up the means is perhaps best illustrated by a numeric example. Suppose you are taking three readings per shift on a process with true mean $\mu = 20$. Table 1.1 shows some hypothetical values that might be obtained. The first column shows the individual readings in time order; the second shows the rational group means that can be calculated after every third original reading. The third and fourth columns are the CUSUM of the individual readings, and that of the rational group means. A glance at the table shows that each of the points of the CUSUM for $\overline{X}$ is three times the corresponding CUSUM for the individual readings. A moment's thought shows that this is not coincidental; after every third observation, the running total of the individual readings must be three times the running total of the rational group means. In general, if the rational group is of size M, then the running total of the individual readings must equal M times the running total of the $\overline{X}$ values. This means that the CUSUM of the $\overline{X}$ values consists of every Mth point from the CUSUM of the individual readings, scaled down by a factor of M.

The CUSUM of the $\overline{X}$ values therefore does not contain any more information than is contained in the CUSUM of the individual readings. On the contrary, it contains *less* information, since it contains only a selection of the points found on the CUSUM of individual readings.

Unlike the case with Shewhart charts, there is nothing to be gained in terms of CUSUM chart performance from using rational group means rather than individual readings. In fact, there is something to be lost by using rational group means rather than individual readings. The point for

Individual reading	Rational group mean	CUSUM of X	CUSUM of $\overline{X}$
23		3	
19		2	
24	22	6	2
16		2	
18		0	
23	19	3	1
15		-2	
14		-8	
19	16	-9	-3
21		-8	
24		-4	
18	21	-6	-2

TABLE 1.1. Comparison of CUSUM of individual observations and CUSUM of rational group means.

the rational group mean cannot be plotted until all the readings for the rational group are in, whereas the CUSUM for individual readings can be updated as each new observation becomes available. The CUSUM of individuals could produce a signal from one of the early observations within the rational group, thereby detecting the shift more rapidly.

This is not to say that CUSUMs of rational group means are always inappropriate. There may be economies of scale that make it a lot easier to sample several items simultaneously than to sample them one at a time, and in this case you should use a CUSUM of the means of these samples. The point being made though is that just as the default for Shewhart charts is rational groups with individual readings as the exception, so in CUSUM charting individual readings are the norm and rational groups are the exception.

1.9 The decision interval form of the CUSUM

The C_n form (or the equivalent S_n form) of the CUSUM is attractive because of the direct interpretation in terms of trendless random walks and linear drifts, but it is not easy to interpret visually. The eye is just not capable of seeing easily whether a V-mask would give a signal, making it necessary to actually use the V-mask to know whether a visually apparent linear segment corresponds to a real shift in mean or is within the realm of random variability. There is another algebraically equivalent form of the CUSUM that is easier to scan visually for evidence of "real" shifts. This is the "decision interval", or DI form of the CUSUM. This is the form we prefer, and we use it for the balance of this work.

At this point, we need to draw a distinction between the CUSUMs that we have been talking about so far, and a new one that we will define. In the remainder of the book, we call a chart using the C_n of Equation 1.1 (or the equivalent S_n) the "V-mask" form of the CUSUM. This name is not a standard one in the CUSUM literature — indeed, there does not seem to be a general consensus on what to call this form.

The DI form of the CUSUM is an algebraic equivalent to setting up the V-mask form CUSUM C_n (or S_n) and diagnosing it using a V-mask of a particular slope k and with particular leg height h. Monitoring X_n for an upward shift in mean is done by setting up the sequence

$$\begin{aligned} C_0^+ &= 0 \\ C_n^+ &= \max(0, C_{n-1}^+ + X_n - \mu - k) \end{aligned}$$

signaling an upward shift in mean if

$$C_n^+ > h$$

If there is a signal, the estimate of m, the time of occurrence of the shift, is given in the C_n CUSUM as that previous value farthest below the V-mask; in the DI form, it is the most recent observation for which $C_m^+ = 0$.

The equivalence of the DI form and the V-mask form using a mask with the same h and k is not too hard to see geometrically or to prove algebraically. Interested readers can see a justification in Van Dobben de Bruyn's 1968 monograph.

E. S. Page (1954) is generally credited with introducing the CUSUM and demonstrating its ability to detect small but persisting shifts. The actual form that he used is a third form, equivalent to the V-mask and the DI forms, but algebraically distinct. His CUSUM was of the form

$$P_0 = 0$$

$$P_n = P_{n-1} + X_n - \mu - k$$

(in other words, the DI form, but without the step of resetting the CUSUM to zero if it should go negative). His rule for a signal was if

$$P_n - \min_{0<m<n} P_m > h.$$

The equivalence of this form to the DI is easy to see.

The DI CUSUM C_n^+ tests for upward mean shifts. Similarly, to check whether there is a downward shift in mean (a previous value sticking out above the upper leg of the mask in the V-mask form) we set up the sequence

$$C_0^- = 0$$
$$C_n^- = \min(0, C_{n-1}^- + X_n - \mu + k)$$

with a signal if

$$C_n^- < -h$$

As with the decision interval CUSUM for upward shifts, when there is an indication of a downward shift in mean, we use the last point m for which $C_m^- = 0$ as the estimate of the instant preceding the change in mean.

The DI form of the CUSUM has a number of practical attractions over the V-mask form. One is that the signal of a shift is very simple: that one or the other of the CUSUMs C_n^+, C_n^- crosses a constant value. This rule is like the familiar Shewhart rule of a point being outside a constant control limit, and is easy to understand and implement. This also makes it easy to see whether a CUSUM is getting close to signaling, and makes it much easier to see where the estimate of the last in-control case following a signal will be.

The DI form also has a practical advantage over the V-mask form of not needing ever-wider graphs to display. Once the CUSUM crosses the decision interval, a signal is generated and presumably the cause of the shift will be diagnosed and the CUSUM restarted.

It is a deficiency of the DI CUSUM that it corresponds to a specific V-mask, or, more accurately, to a V-mask with a specific slope k. In the V-mask CUSUM, it is possible to overlay the CUSUM with different V-masks (and there may be circumstances in which this makes sense), but a change in k can only be addressed in the DI format by constructing a new CUSUM. This is easily done if one is using a spreadsheet to obtain the CUSUM. (We include some Excel CUSUM files on the Web site that allow easy changes in k.)

We can use the decision interval form of the CUSUM to estimate the magnitude of the shift in mean much as we used the C_n form. The segment of the decision interval CUSUM leading to the signal starts with some case m for which $C_m = 0$ and then is positive up to the point at which it crosses the decision interval h. This shows that in this segment

$$C_n^+ = C_m^+ + \sum_{i=m+1}^{n} (X_i - \mu - k).$$

Following the shift in mean from μ to $\mu + \delta$, the summand has a normal distribution with mean $\delta - k$. This means that we can estimate δ by adding k to the slope of the decision interval CUSUM from point m to point n, getting the estimate:

$$\hat{\delta} = k + \frac{C_n^+ - C_m^+}{n - m} \tag{1.5}$$

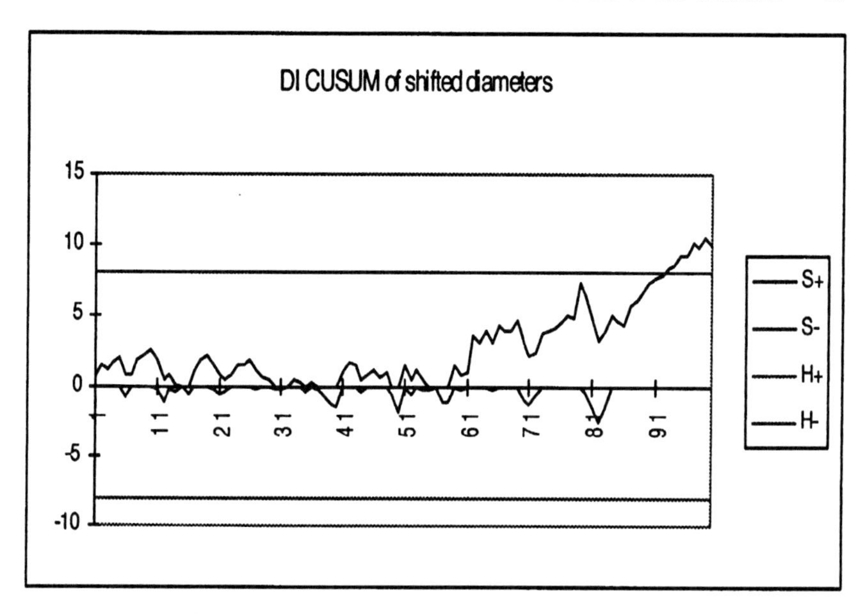

FIGURE 1.10. Decision interval form of the CUSUM.

which (because $C_m^+ = 0$) reduces to $k + C_n^+/(n-m)$. This is the same estimator that we got by looking at the slope of C_n form of the CUSUM.

Similarly, if the downward decision interval CUSUM signals a shift, then the magnitude of the shift can be estimated by

$$\hat{\delta} = -k + C_n^-/(n-m).$$

Note that since C_n^+ is necessarily positive, and C_n^- is necessarily negative, it is impossible for the estimate of δ to lie between $-k$ and k. This is another reminder of the fact that the estimate of the shift produced by the CUSUM is biased away from zero.

The CUSUMs C_n^+ and C_n^- are decision interval equivalents for the unstandardized V-mask-type CUSUM C_n. Similarly, the standardized V-mask-type CUSUM S_n has decision interval equivalents S_n^+ and S_n^- defined by the recursive equations

$$\begin{aligned} S_0^+ &= 0 && (1.6)\\ S_0^- &= 0 && (1.7)\\ S_n^+ &= \max(0, S_{n-1}^+ + U_n - k) && (1.8)\\ S_n^- &= \min(0, S_{n-1}^- + U_n + k). && (1.9) \end{aligned}$$

These standardized forms have the same advantages and disadvantages over the unstandardized forms as does S_n over C_n.

1.9.1 Example

We close this discussion with a look at Figure 1.10, which is the decision interval form of the CUSUM of our shifted bolt diameter example.

This CUSUM shows the upward CUSUM S_n^+ and the downward CUSUM S_n^- along with decision intervals at $h = 8$ and $h = -8$ for diagnosing them. Although the downward CUSUM remains close to the axis for the whole sequence, your eye is drawn to the upward CUSUM. After meandering along for the first half of the sequence, it rose firmly after about point 50, breaking out of the decision interval shortly after point 90. This signals the presence of the shift. If we want to estimate its magnitude, we could read the chart more carefully, or go back to the actual tables of the numbers being plotted. These tables show that the last point for which S_m^+ was zero was point 58, and the breakout above $h = 8$ occurred on point 93, where $S_n^+ = 8.37$. As we are using $k = 0.25$, this gives the estimated shift as

$$\hat{\delta} = 0.25 + 8.37/(93 - 58) = 0.49$$

standard deviations, or (going back to the original units) 0.049mm. Not coincidentally (since they are different ways of implementing the same technique) this diagnosis is exactly the same as that given earlier using the V-mask form of the CUSUM.

Finally, we have defined the downward CUSUM so that its value is always non-positive. Some workers prefer DI CUSUMs that are always nonnegative, and use the definition

$$\begin{aligned} S_0^- &= 0 \\ S_n^- &= \max(0, S_{n-1}^- - U_n - k) \end{aligned}$$

This choice works just as well as ours, and the worst that can be said for there being two commonly used conventions is the possibility of confusion.

There is no profound theoretical reason to prefer defining S^- to have non-positive values (as we do) rather than nonnegative (as some others do); we prefer negative values since a downward mean is signaled by a downward breakout of the pair of CUSUMs S_n^+, S_n^-. It seems more natural to diagnose a downward shift by a large negative value than by a large positive one.

Notice that it still took about 40 observations for the change to be signaled. We show later that the CUSUM gives the fastest possible detection time, on average, for persistent shifts. That fastest possible time can still be fairly long for small shifts, which are difficult to detect using even the best methodology. This example illustrates that point.

1.10 Summary

Unlike Shewhart control charts, CUSUM control charts are designed to detect persistent shifts away from the in–control distribution of a process. These persistent shifts are best detected by CUSUM charts.

In this introductory chapter, we have introduced the V–mask and decision interval form of the CUSUM control chart. Both work best with individual observations, unless there are compelling economic or process reasons to work with rational groups. CUSUM charts allow the prompt detection of out–of–control states, and provide estimates of when the change occurred, and of the magnitude of the change.

We illustrated with charts for detecting mean shifts for a normal process. We derive CUSUM schemes for the key members of the exponential families in subsequent chapters.

In the next chapter, we discuss the design of a CUSUM chart. This design involves selecting k and h so that the resulting chart has "good" properties.

1.11 Further reading

Walter Shewhart developed the original "Shewhart" scheme, and his book *Economic Control of Quality of Manufactured Product* is the classic work on Shewhart Charts. A very accessible discussion of the Shewhart scheme can be found in Montgomery's text on quality control, *Introduction to Statistical Process Control.*

The "Western Electric" zone rules were an early attempt to improve the response of Shewhart charts to moderate shifts. They signal if successive points on a Shewhart chart fall within particular ranges. Champ and Woodall (1987) confirmed that the best of these zone rules improve the Shewhart chart considerably, but do not bring its performance up to CUSUM levels.

Page's (1954) paper "Continuous inspection schemes" introduced the idea behind cumulative sum control charts, although not its present form. The work was extended in the late 1950s and early 1960s by many statisticians. Van Dobben de Bruyn's 1968 monograph is an excellent summary of much of this work.

The trilogy by Johnson and Leone (1962) provided a compact exposition of the theory underlying CUSUMs, and developed the optimal CUSUM schemes for a number of the distributions we cover in detail.

Most standard texts on statistical process control give at least some discussion of CUSUMs; Montgomery is a good source relating CUSUM ideas to other SPC methodologies. Goel (1982) provided a brief sketch of CUSUMS, concentrating on control of a normal mean.

The CUSUM accumulates information on a number of cases in a window of varying width. The "moving sum" or MOSUM approach uses the total (or equivalently the average) of a fixed number of observation: as each new reading is added on the right of the window, the oldest reading is removed on the left. See, for example, Bauer and Hackl (1978, 1980) for a discussion of industrial application of MOSUMs for location and scale control. Although MOSUMs share the CUSUM's intuitive appeal of accumulating information across readings, they do not have matching optimality properties.

Two companion papers by Bissell (1984) showed how an out-of-control mean varying linearly with time leads to a parabolic expected CUSUM. He showed that the performance of the CUSUM for detecting this shift is respectable, if not stellar, and proposed its use to estimate the constant of proportionality. Gan (1992) used a modified Markov chain approach to find accurate run lengths in this linear drift model.

2
CUSUM design

> Adding one thing to another to discover the scheme of things...
> *Ecclesiastes 7:27*

2.1 The choice of k and h

In Chapter 1, we illustrated the major properties of the CUSUM: its descriptive properties, its ability to signal persistent shifts, even if these are quite modest, and its diagnosis using either the V-mask or the equivalent decision interval form. In this chapter, we look more closely at the design of the CUSUM – that is, at the choice of the parameters k and h that define the decision interval scheme or the equivalent V–mask .

Throughout this chapter, we work with the CUSUM of standardized data, obtained by first transforming the process readings by subtracting their in-control mean and dividing by their standard deviation. This assumption is made to simplify the notation and discussion a bit. It does not detract from the generality of the CUSUM, since the CUSUM of the data on the original scale is just the CUSUM on the standardized scale, multiplied by the scale factor σ. Since the CUSUMs are assumed standardized, the chart constants k and h are dimensionless; they are in units of the standard deviation of the original process measurements.

The discussion continues the framework used in Chapter 1, controlling the mean of $N(\mu, \sigma^2)$ quantities. The use of this framework here does not limit the discussion. The general principles hold whether we are monitoring normal data or data from any other distribution, discrete or continuous. It is just in the details that the distribution of the readings becomes important, and we cover those details in later chapters.

2.1.1 Reference value k - "tuning" for a specific shift

The chart constant k is called the "reference value" or the "allowance". The chart constant h is called the "decision interval." In Chapter 1, we introduced the parameter k by suggesting that it be one-half Δ, the anticipated shift in the process mean. This implies that we may have the same stream of process measures, but would pick different chart constants if we changed our minds about the likely size of the shift in process mean. The idea of having different chart constants for different circumstances is perhaps novel to those trained in Shewhart chart methodology (where it is almost universal to put the control limits at three standard errors from the mean), but is an important feature of CUSUM chart design.

As we show in the more detailed theoretical discussion in Chapter 6, this intuitive choice for the reference value k is the right one for the shift in mean of a normal distribution; choosing a particular value of k implies that the CUSUM is being designed to detect a shift of $\Delta = 2k$ standard deviations.

There are two routes by which one can come to a good choice of Δ. First, there are some circumstances in which, when the process goes out of control, there is one particular level to which it is likely to move. As an example, consider monitoring the emissions from a gasoline engine to check whether all cylinders are firing properly. If one of the cylinders is not firing (perhaps having no spark), then its fuel would be emitted unburned. This means that, if the process were to go from the in-control state with all cylinders firing to an out-of-control state, the most likely out-of-control state would be one in which one cylinder was not firing. With theoretical calculations or previous experimental evidence, you might know what the mean emission would be with one cylinder not firing, and then the CUSUM could sensibly be tuned to this choice of Δ.

The other and more common possibility is that you choose the size of shift for which you desire the quickest detection. This would be a shift large enough to have a meaningful impact on the process operation but small enough not to be obvious to the naked eye. The trade-off here is that the scheme that is optimal for detecting a shift of three standard deviations may give you a scheme that is not very good at detecting a shift of one standard deviation. If a shift of one standard deviation is too small to bother you, then that trade-off is fine. If however a shift of one standard deviation is large enough to affect the process, then not being able to detect it reliably could be a problem, and then tuning to $\Delta = 3$ could be a bad choice.

Aiming at a too-small shift is also potentially harmful. It is possible to design CUSUMs for the detection of tiny shifts in mean such as one quarter of a standard deviation. However, these CUSUMs do not respond quickly to shifts of *almost any* magnitude, even large ones.

> k is chosen for optimal response to a shift of a specified size Δ. It is not optimal for shifts of substantially different size.

2.2 Runs, run length, and average run length

The CUSUM chart starts out at its initial state S_0^+. From there, it may stay on the axis, or it may move into positive values. Each episode of positive values will end in one of two ways: either the CUSUM returns to zero, or it crosses the decision interval. When the chart crosses the decision interval, this indicates that a shift has occurred and actions will be taken to diagnose the shift. Generally the CUSUM will then be restarted. The whole sequence going from the starting point to the CUSUM crossing the decision interval is called a *run*. The number of observations from the starting point up to the point at which the decision interval is crossed is called the *run length* .

Sometimes the CUSUM will signal when in fact no shift has occurred. This false alarm is analogous to a Type I error in classical hypothesis testing. These false alarms are undesirable, as they cause us to waste time and energy and disrupt operations looking for nonexistent special causes. We would like the runs between the inevitable false alarms to be as long as possible.

A Type II error in classical hypothesis testing also has an analog in control charting — this is a chart remaining within its decision interval even though a special cause has surfaced. If there has been a shift big enough to have practical implications, you would like to detect it as soon as possible. These objectives (of long runs before false alarms but short runs before the chart signals actual shifts) conflict, so it is necessary to make trade-offs between them. This is, of course, exactly analogous to the trade-offs between Type I and Type II errors in classical hypothesis testing.

The run length is a random variable, having a mean, a variance and a distribution. Its mean is called the *average run length* or *ARL*. The ARL is an imperfect but useful summary number of the general tendency toward long or short runs. It is less than perfect since the run length distribution turns out to be highly variable. A high in-control ARL does not rule out the possibility of a very short run before the CUSUM gives a false alarm, and a low out-of-control ARL for the CUSUM is no guarantee that there would not be a long run before the CUSUM detected an actual shift. Despite these imperfections, the ARL is an easily interpreted, well-defined measure for which we now have quite good algorithms. It is the standard measure of performance of the CUSUM.

The in-control behavior of the CUSUM is measured by its in-control ARL. We want this to be sufficiently long. The out-of-control ARL is the ARL of the CUSUM following a shift in the process mean. We want this to be as short as possible.

There is some indeterminacy in defining the ARL. When we start the CUSUM, it is commonly at $S_0^+ = 0$. The ARL corresponding to this state is the average number of observations that will elapse from the startup to the first false alarm. In fact, every time the CUSUM goes back to zero, it loses its memory of its earlier history, so this ARL is also the ARL of the CUSUM, not just from when the CUSUM was first started, but from any time that it has come back to its starting value.

We could imagine a different ARL. Suppose we allow the CUSUM to run for a long time, restarting it any time it goes beyond its decision interval. Then its value will settle down to a "steady state"; for some fraction of the time it will be zero, and otherwise it will have a value following some statistical distribution taking on values between 0 and h. This distribution is termed the *steady-state distribution* of the CUSUM. We could then start the clock at some arbitrary moment when the CUSUM had attained its steady state distribution and measure the ARL from that instant up to the next signal; this would give us the *steady-state* ARL.

The steady-state ARL is necessarily smaller than the ARL from startup since the CUSUM will be starting the run from a value that could already be on the way to h.

A similar question arises when we think about the ARL when the process undergoes a shift. We can suppose that the shift occurs when the CUSUM starts (so that the CUSUM is starting out from zero), or we can suppose that the shift occurs at some instant when the CUSUM has been running in control for long enough to reach its in-control steady-state distribution. Here too, the ARL from the steady-state will be less than the ARL starting out from zero.

It is most common to use ARLs starting from S_0^+ rather than from the steady state, and our discussion of ARLs concentrates on this approach. Although the ARL starting from the steady state will be smaller than that starting from zero, the relative rankings of different CUSUMs are generally not greatly affected by the assumption about the starting point.

> The ARLs in control and out of control are the performance measures for a CUSUM. We use the ARL from S_0^+.

2.2.1 *The choice of h, the decision interval*

The in-control ARL of the CUSUM depends on the values of k and h. Larger values of either of these parameters leads to larger ARLs. At the extremes, if h and k are both zero, then the ARL is 1, since the first point will necessarily give a value of at least 0. The ARL can obviously be made as large as you like by increasing k or h or both.

We mentioned earlier that the choice of k follows from the choice of the shift in mean to which the CUSUM is to be tuned. The other parameter, the decision interval h, is usually fixed by deciding on the minimum tolerable in-control ARL. When X is a continuous random variable, whatever value is given to k, it is possible to find some h such that the ARL will equal any given target greater than 1.

Thus for a given distribution of the process measurement X, choosing k and h implies an ARL, and choosing k and any ARL greater than 1 implies an h. This is true regardless of the distribution of X, although the relationship between k, h, and the ARL depends crucially on the distribution of X.

If X is discrete (a situation that is covered in detail in Chapter 5), this general claim that we can always find an h that gives a target ARL needs to be modified slightly. For example, if X is an integer-valued discrete random variable and k is also an integer, then only integer values of S_n^+ are possible. This means that the ARL will be a step function of h, increasing as h passes through integer values but remaining constant as the fractional part of h changes. Similarly if X is integer and k is a half-integer, then the attainable values of S_n are integers or half-integers. This means that choosing k and a target ARL will not in general lead to a value of h that exactly attains the target ARL. It will, however, be possible to select an h whose ARL is at least the target ARL.

- Given h and k, we can find the *ARL*.
- Given *ARL* and k, we can find h.

2.2.2 *Calculating the k, h, ARL relationship*

The CUSUM is designed by choosing k and h values that give some acceptably high target value for the in-control ARL starting out from S_0^+. This can be done using tables, graphs, or software.

Tables, such as Table 3.1, showing the ARL for selected values of k and h may be scanned to find the h that gives an acceptable ARL. For design purposes, tables that show the required h for selected values of k and the ARL are better. Table 3.2 is an example of this type of table.

Suppose you want to set up a CUSUM scheme to detect a shift of 1 standard deviation as quickly as possible. This target shift corresponds to a k of 0.5. If an in-control ARL of 500 is the highest false alarm rate you are willing to consider, then Table 3.2 shows that the CUSUM should have $h = 4.389$. At this same target shift, if you wanted to decrease the false alarm rate by pushing the in-control ARL out to 1,000 readings, then the required h would become $h = 5.071$.

Both types of table are inherently less flexible than computer programs that calculate the ARL when provided with k and h. The programs integrated with this book provide an easy way of calculating the ARL for particular choices of these quantities and, from that ARL, for finding a suitable CUSUM.

The code ANYARL is a software alternative to Table 3.1 of Chapter 3. When supplied with a (k, h) pair, it calculates the ARL.

The program ANYGETH is a software alternative to Table 3.2. When supplied with a (k, ARL) pair, it calculates the h value.

Generally, these software approaches provide the preferred way of calibrating CUSUMs, with printed tables being most useful for quick, approximate, or exploratory calculations.

A full CUSUM design involves two parts. First there is the problem of selecting a k, h pair that will provide acceptable in-control ARL. The second part is no less important. It involves assessing how quickly the CUSUM will respond to out-of-control situations when the parameter value has shifted. As we will see later, considerable progress in calculating the out-of-control ARL can be made using just in-control calculations. Tables are considerably less satisfactory than software for finding the ARL for the out-of-control state.

2.2.3 A closer look at the choice of in-control ARL

We talk about the ARL measured in terms of process readings, but what really matters is the time interval between false alarms. For example, if a three-shift process gathers four readings per shift, this gives around 350 readings per month. If it is considered acceptable to have one false alarm per month, then the in-control ARL should be set to around 350.

Consider again the bolt diameter data used in Chapter 1. These diameters had an in-control true mean of $\mu = 5$mm, and a true standard deviation of $\sigma = 0.1$mm. Let's design a CUSUM tuned to a shift of 0.1mm. This is 2% of the mean diameter, and is equal to σ. When we standardize the data by subtracting μ and dividing by σ, then the shift to which we want to tune the CUSUM will be of size $\Delta = 1$.

This implies the choice of reference value $k = 0.5$. We illustrate the design in Figure 2.1 with some calculations using GETH, the program from the Web site for finding the h that gives a particular target ARL for a given k.

```
Program to calculate CUSUM decision intervals
Copyright 1997, D M Hawkins and D H Olwell
Which distribution do you want? (Give its number from
this menu:)
1 Normal location
2 Normal variance up
3 Normal variance down
4 Poisson up
5 Poisson down
6 Binomial up
7 Binomial down
8 Neg binomial up
9 Neg binomial down
1

What is the Winsorizing constant?
(enter 1000 if you don't want to winsorize or don't
understand the question)
1000

Do you want zero-start (say Z) or FIR (say F)?
Z

k .5000 h 4.0410 ARL 349.84
```

FIGURE 2.1. Output from ANYGETH

The choice of decision interval $h = 4.041$ gives the target in-control ARL of 350.

Suppose now that the same operator is maintaining four machines with CUSUM control on each. If a total of one false alarm per month can be tolerated, then the false alarm rate per machine goes out to four months, or around 1,500 process readings. The in-control ARL of each chart should then be set by suitable choice of k and h to about 1,500. Redesigning the CUSUM to have an in-control ARL of 1,500 readings can then be done with the following command to ANYGETH.

```
Enter k and ARL values (zeroes stops run)
.5 1500

k .5000 h 5.4727 ARL 1501.12
```

The h value needed to get an in-control ARL of 1,500 with $k = 0.5$ is then $h = 5.473$.

Going back to the single machine, if we take 1 reading every 30 minutes rather than every 2 hours, then 1 month's data amount to around 1,500 readings, and so to keep the false alarm rate down to 1 per month we would need the in-control ARL to be 1,500. Putting the scenarios together, with 4 machines all on a 30-minute reading cycle, keeping the chart down to a total of 1 false alarm per month would involve setting the in-control ARL of each chart to 6,000. GETH gives the output

```
Enter k and ARL values (zeroes stops run)
.5 6000

k .5000 h 6.8516 ARL 6003.00
```

increasing the needed h to 6.852.

As a final concern, we will likely want to be checking for both increases and decreases in the mean. This means that to keep the overall in-control ARL down to one signal per month, the ARLs that we sketched all need to be doubled. The ANYGETH output reflecting this fact and getting the required h values are

```
Enter k and ARL values (zeroes stops run)
.5 700

k .5000 h 4.7188 ARL 699.70

Enter k and ARL values (zeroes stops run)
.5 3000

k .5000 h 6.1602 ARL 2999.10

Enter k and ARL values (zeroes stops run)
.5 12000

k .5000 h 7.5430 ARL 12001.79

Enter k and ARL (zeroes stops run)
0 0
Would you like to run some other distribution or
settings?
n
Stop - Program terminated.
```

As this discussion has tried to emphasize, there is no single in-control ARL that is suitable for all purposes. Instead, the in-control ARL should be chosen in a context-specific way to control the frequency of false alarms and to ensure adequate response times to genuine shifts. Papers about CUSUM charts often use some standard in-control ARL such as 500 or 1,000, and in this book we tend to do so also. This is primarily for the convenience of working with some particular ARL. For actual use of the chart, it is important to realize that it is not necessary to accept some standard off-the-shelf in-control ARL; the sampling times and the CUSUM chart parameters can be set to achieve a good combination of in-control and out-of-control behavior.

Shewhart charts usually follow the default setting of control limits placed three standard errors above and below the mean of the quantity being monitored. For the Xbar chart for a normal process, this corresponds to an in-control ARL of 370 rational groups. For a process producing one rational group per shift, this would correspond to a false alarm every four months for any single chart. Using the default control limits leads to the overall false alarm rate going up in direct proportion to the number of charts being maintained, a situation that might be undesirable where there were limited resources for tracking down chart signals. The idea of setting the control limits to fix the overall false alarm rate of the control charts as a group can, of course, also be implemented with Shewhart charts by setting the control limits at suitable multiples of the standard error. This is a good idea where you want to maintain control of the overall false alarm rate.

> Choose an ARL tailored to your needs that reflects the total rate of false alarms you can tolerate.

2.2.4 Designing a CUSUM of Xbar

As mentioned in Chapter 1, CUSUMs are usually made of individual readings rather than rational group means. This is because using means rather than individual readings does not improve performance (as it does with Shewhart charts) but even degrades performance slightly. Still, there is nothing to stop us using rational group means if they make sense on some other grounds — for example, labor savings. Let's return to the bolt diameter problem and suppose that for administrative reasons we wish to take a rational group of size $m = 5$ each shift and retain the false-alarm rate at 1 per month. This involves an ARL of roughly 100 rational groups.

When we come to k, some care is needed. We want to tune the CUSUM for a shift of 0.1mm in the mean. This corresponds to $\Delta = 1$ standard deviation of the bolt diameter. When we are working with the means of rational groups of size 5 though, the standard error of the mean is $\sigma/\sqrt{n} = 0.1/\sqrt{5} = 0.0447$, and so a shift of 0.1mm corresponds to $0.1/0.0447 = 2.236$ standard errors of the mean. This value of $\Delta = 2.236$ then leads to the choice $k = 1.118$. Our problem then becomes one of finding the h that, with $k = 1.118$ will have an in-control ARL of 100. Running this pair through ANYGETH then gives

```
Enter k and ARL values (zeroes stops run)
1.118 100

k 1.1180 h 1.3467 ARL 99.96
```

So the rational groups of size 5 will be reduced to their means $\overline{X}_n$. These will be standardized to form

$$U_n = \frac{\overline{X}_n - 5}{0.1/\sqrt{5}}.$$

These standardized values will be CUSUMed using the reference value $k = 1.118$:

$$S_0^+ = 0$$

$$S_n^+ = \max(0, S_{n-1}^+ + U_n - 1.118)$$

and the CUSUM will signal if S_n^+ exceeds 1.347.

If we prefer to use the original scale, this CUSUM can be rescaled back to millimeter units by multiplying k and h by 0.0447, the standard error of $\overline{X}$, resulting in:

$$\begin{aligned} C_0^+ &= 0 \\ C_n^+ &= \max(0, C_{n-1}^+ + (\overline{X}_n - 5) - 0.05) \end{aligned}$$

with a signal if $C_n^+ > 0.060$.

On the original scale, the reference value that is subtracted from each $\overline{X}$ is 5.05, a value midway between the in-control mean of 5 and the out-of-control level 5.1 to which the CUSUM is tuned. This is what you should expect, since the CUSUM is still tuned to detect a shift from 5.0 to 5.1mm, though the route by which we got there with the factor $\sqrt{5}$ being introduced and then canceled out may obscure this.

The S^+ and C^+ charts are identical apart from the scaling on their vertical axes.

Earlier, we showed by example the close connection between a CUSUM of means and one of the original individual readings. It is interesting to do a further rework of this original-scale CUSUM of the $\overline{X}$ values, rewriting it in terms of the original observations. Let's temporarily write X_{nj} for the jth observation in the nth rational group, so that

$$\overline{X}_n = \frac{\sum_{j=1}^{m} X_{nj}}{m}.$$

Then multiplying C_n^+ by m gives the recursion

$$m\, C_n^+ = \max\{0, m\, C_{n-1}^+ + \sum_1^m (X_{nj} - 5.05)\}$$

and signals if

$$mC_n^+ > mh = 0.30.$$

The CUSUM of individual readings X_{nj} transforms back to the original scale[1] as

$$C_{n,j}^+ = \max\{0, C_{n,j-1}^+ + (X_{nj} - 5.05).\}$$

To match the false alarm rate of once per 100 rational groups, on the individual-observation CUSUM we would need an in-control ARL of 5×

[1] As a matter of notation, we need to interpret the subscript $\{n, 0\}$ as $\{n-1, m\}$ to get the ordering we want.

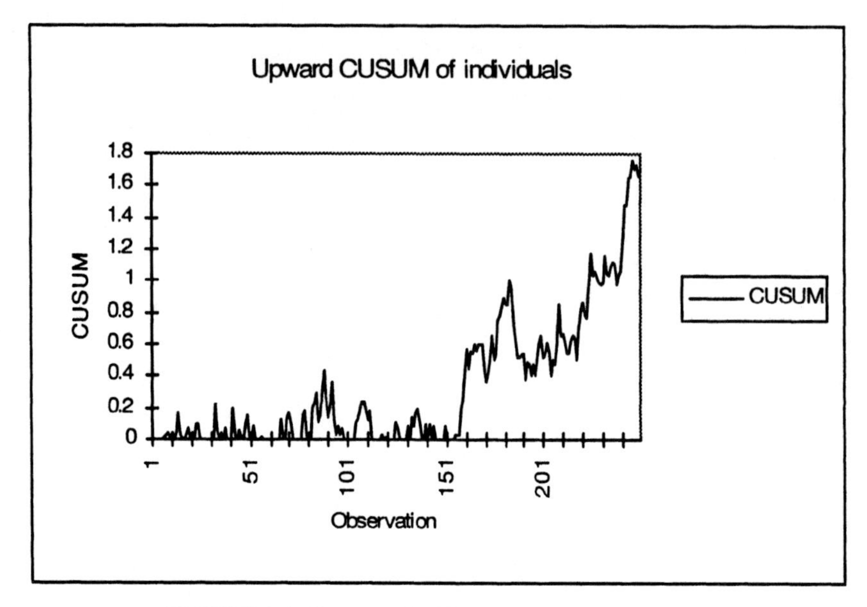

FIGURE 2.2. Upward CUSUM of individual readings.

100 = 500 readings. Then ANYGETH shows that $h = 4.389$ standard deviations will give this in-control ARL, so the CUSUM of individuals will signal if

$$C_{n,j}^{+} > 0.1 \times 4.389 = 0.44$$

There is a close connection between these two CUSUMs. Ignoring the distraction of the subscripts on X, each accumulates the deviation of the individual observations from the reference value 5.05. The CUSUM of individual observations resets the cumulative sum to zero whenever any partial sum takes it below zero; the CUSUM of rational means does so only at the end of the rational group. It follows then that the CUSUM of individual readings will always be at least as large as that of the rational group means. This is the reason for its needing a larger decision interval to attain a comparable in-control ARL.

Figures 2.2 and 2.3 make this comparison. They show data CUSUMmed, first as individual readings, and then as means of rational groups. The sequence is in control for the first 60% of its length, but then undergoes a change. Both CUSUMs clearly show the upward drift following the change, but they differ in the look of the in-control portion. The CUSUM of individual readings shows several excursions above the axis, but the CUSUM of the means is mainly on the axis. The reason for this is that the reference value is quite likely to be exceeded by an individual reading as it is only one half of a standard deviation away from the mean. The reference value for the CUSUM of means though is 1.118 standard deviations away from the mean and is seldom exceeded. Following the change, both CUSUMs have

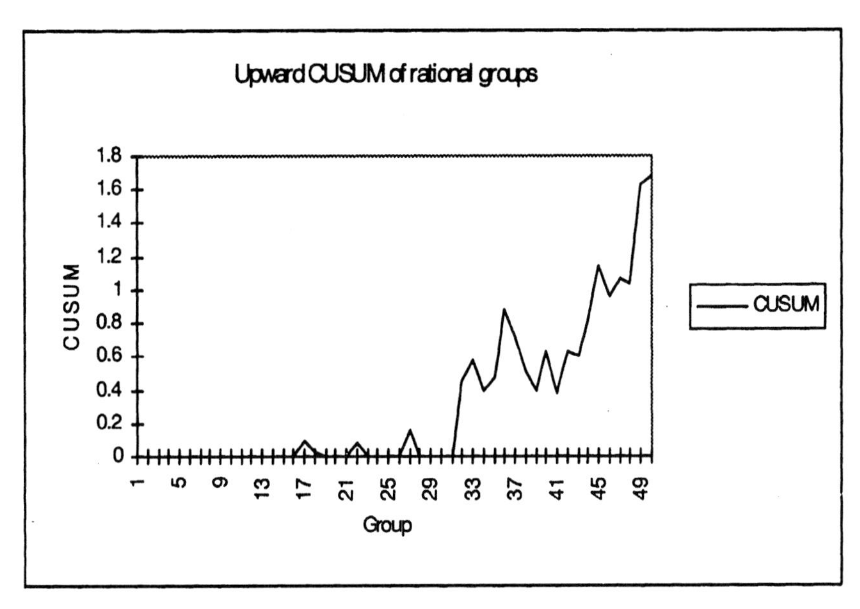

FIGURE 2.3. Upward CUSUM of means of rational groups of size 5. Compare with Figure 2.2.

very similar shapes, although the Xbar CUSUM is naturally smoother than the individual reading CUSUM.

> Design of an Xbar CUSUM chart can be accomplished easily by standardizing and using the same approach as for a CUSUM of individual observations.

2.3 The Shewhart Xbar chart as CUSUM

The standard Shewhart Xbar chart breaks the readings down into rational groups of size m say. A signal of upward shift is then given if $U_n > 3$, where

$$U_n = \frac{(\overline{X}_n - \mu)}{\sigma/\sqrt{m}}.$$

It is easy to see that this rule is identical to the rule

$$\begin{aligned} S_0^+ &= 0 \\ S_n^+ &= \max(0, S_{n-1}^+ + U_n - 3) \end{aligned}$$

with a signal if $S_n^+ > 0$. This means that the Shewhart chart is the same as a CUSUM of rational group means, using $k = 3$ and $h = 0$.

This shows that the Shewhart chart with its control limits set at the conventional 3 standard errors can be thought of as a CUSUM "tuned" to a shift in mean of 6 standard errors, and with its in-control ARL set to 740.

The connection between the Shewhart chart and the superficially very different CUSUM chart is instructive. It helps, first, to illustrate the intuitive idea that the Shewhart chart is very effective for detecting large shifts in mean. In particular, the I chart (a Shewhart chart of individual observations) is the "best" possible chart for detecting shifts in mean of magnitude 6 standard deviations.

The connection may also be of some help in selecting the value of m in a Shewhart chart. Suppose, for example, that you want to select the rational group size for a Shewhart Xbar chart that would be suitable for detecting shifts of, say, 2.5σ. Equating this to 6 standard errors then gives the equation

$$2.5\sqrt{m} = 6$$

giving $m = 5.76$, a size close to the value 4 or 5 so often used for rational groups in Shewhart charting.

Reversing this reasoning, a Shewhart Xbar chart using rational groups of size 5 can be thought of as a CUSUM tuned to a shift of $6/\sqrt{5} = 2.7$ standard deviations. For shifts of about this magnitude, the conventional Shewhart chart has the same optimality as the CUSUM.

2.4 Summary

The average run length ARL is the usual performance measure for CUSUM charts.

To design a CUSUM scheme, we first select the out–of–control state for which we would like maximum sensitivity. This determines the reference value k. We then select the desired in–control ARL to meet our particular needs. Once the ARL and k are determined, the value of the decision interval h, follows, and may be found from tables or software.

In the next chapter, we continue our examination of CUSUMs of normal data.

2.5 Further reading

We have concentrated on designing by choosing the in-control and out-of-control means. Bissell's (1969) discussion paper had a perspective largely lost from more recent writings, connecting the CUSUM design with acceptance sampling ideas of acceptable and rejectable quality levels. He also showed how to handle the effect of variance changes using in-control calculations (specifically, an ARL nomogram). We discuss this topic in Chapter 3.

The V-mask is a geometric equivalent to the decision interval form that is optimal for a step shift to some known out-of-control level. If the mean shifts to some different level, then the performance may suffer. Two proposals to improve the response to a wider range of out-of-control levels are to replace the standard V-mask with some other shape. Bissell (1979) proposed a parabola, and Rowlands et al (1982) suggested a snub-nosed mask. In each case, the effect of the change is to cut away the sharp corners of the V-mask, giving a mask that will respond more quickly to large changes.

As we stressed, the choice of h is made indirectly as a consequence of choosing an in-control ARL. In some circumstances, you can attach quantifiable economic costs to running an out-of-control process, to stopping and diagnosing an in-control process, and to the sampling needed for SPC. When this is the case and you specify some probabilities associated with going out of control, then it becomes possible to set up the CUSUM to minimize the expected total cost. Bather (1963), Chiu (1974), and Taylor (1968) discussed incorporating these economic considerations in chart design.

The run length is highly variable, and the ARL is a necessarily crude summary that may be supplemented with the standard deviation of the run length (SDRL)

The concern about possible long runs can be addressed through bounds (Waldmann, 1986) and approximations (Woodall, 1983) on the probabilities of run lengths. A more comprehensive analysis is through the full distribution of the run length. The run length of a Shewhart chart follows a geometric distribution (and so has a standard deviation one less than the ARL). The geometric distribution is a reasonable approximation for the run length of CUSUMs with large k values, but smaller k values lead to somewhat less heavy-tailed run length distributions. Gan's (1993b) program provides this for a CUSUM of normal data, and that of Gan (1992a) for exponential data

In the book, we work with run lengths measured in numbers of observations. This is equivalent to measuring in elapsed time, provided the readings are equally spaced in time. In "variable sampling intervals" (VSI) methods, this is not the case, and sampling frequency is increased when there are early indications of a possible out-of-control, and decreased when these indications go away. Reynolds et al. (1990) showed how the use of VSI with CUSUMs can improve their performance. A variant that combines the CUSUM with a short-term SPRT is given in Assaf and Ritov (1989).

3
More about normal data

All models are wrong, but some are useful.
G. E. P. Box

The most common application of CUSUMs is controlling the mean of a stream of process data modeled by the normal distribution. The last two chapters illustrated the CUSUM for the mean of normal data; this chapter gives more detail. We look at extensions to the basic method, covering Fast Initial Response CUSUMs, CUSUMs for variance shifts, weighted CUSUMs, and combined Shewhart/CUSUM charting. We also examine the effects of model departures on CUSUM schemes. We begin with a closer look at the mechanics of determining ARLs.

3.1 In–control ARLs

We start out with two ARL tables. These tables were computed using the same computer codes as are used in the ANYARL and ANYGETH procedures mentioned in Chapter 2. It is not always convenient to use a computer, particularly for quick or approximate design problems, and Tables 3.1 and 3.2 are useful paper tools for calculations on CUSUMs for the mean of normal data. Table 3.1 lists the ARL of a CUSUM of standardized data for a wide range of k and h values. Entries in which the ARL is in excess of 100,000 are generally not of interest, and so are omitted.

k	-0.25	0.00	0.25	0.50	0.75	1.00	1.25	1.50
h								
1.000	3.43	4.75	7.0	11.2	19.2	35.3	68.9	142.2
1.125	3.75	5.27	8.0	13.2	23.4	44.8	91.4	196.8
1.250	4.08	5.84	9.1	15.4	28.6	57.2	122.1	274.9
1.375	4.42	6.44	10.3	18.0	34.9	73.1	164.0	387.2
1.500	4.78	7.09	11.6	21.1	42.6	93.8	221.5	549.7
1.625	5.15	7.76	13.0	24.6	52.0	120.7	300.5	786.0
1.750	5.53	8.48	14.6	28.6	63.5	155.5	409.4	1130.8
1.875	5.92	9.22	16.3	33.3	77.4	200.5	559.4	1635.8
2.000	6.32	10.00	18.2	38.5	94.3	258.7	766.2	2376.8
2.125	6.72	10.81	20.2	44.6	114.9	333.8	1051.0	3465.4
2.250	7.13	11.66	22.4	51.5	139.7	430.7	1443.0	5065.1
2.375	7.54	12.53	24.7	59.3	169.7	555.5	1981.9	7414.5
2.500	7.96	13.43	27.3	68.2	206.0	716.0	2721.5	10861.4
2.625	8.39	14.37	30.0	78.3	249.7	922.2	3735.3	15910.5
2.750	8.81	15.33	32.9	89.8	302.5	1187.0	5123.1	23294.0
2.875	9.24	16.32	36.1	102.8	366.1	1526.8	7020.6	34071.6
3.000	9.68	17.35	39.5	117.6	442.8	1962.8	9613.2	49777.5
3.125	10.12	18.41	43.1	134.4	535.3	2522.1	13153.3	72633.4
3.250	10.56	19.50	47.0	153.4	646.9	3239.6	17985.8	
3.375	11.01	20.61	51.2	175.0	781.4	4160.1	24582.0	
3.500	11.46	21.76	55.7	199.6	943.7	5341.4	33586.2	
3.625	11.91	22.95	60.5	227.4	1139.4	6857.6	45879.9	
3.750	12.37	24.16	65.7	259.0	1375.6	8803.9	62669.1	
3.875	12.83	25.40	71.2	294.8	1660.5	11302.8	85603.0	
4.000	13.29	26.68	77.1	335.4	2004.2	14511.5		
4.125	13.75	27.99	83.4	381.4	2418.9	18631.7		
4.250	14.22	29.32	90.2	433.6	2919.2	23922.8		
4.375	14.68	30.69	97.4	492.8	3522.7	30717.7		
4.500	15.15	32.09	105.1	559.9	4250.8	39443.6		
4.625	15.62	33.53	113.4	636.0	5129.1	50649.1		
4.750	16.10	34.99	122.2	722.3	6188.6	65038.7		
4.875	16.57	36.48	131.6	820.1	7466.6	83516.4		
5.000	17.05	38.01	141.7	930.9	9008.2			
5.125	17.53	39.57	152.4	1056.5	10867.8			
5.250	18.01	41.15	163.9	1198.9	13111.0			
5.375	18.49	42.77	176.1	1360.2	15816.7			
5.500	18.97	44.42	189.2	1543.1	19080.5			
5.625	19.45	46.11	203.2	1750.4	23017.3			
5.750	19.93	47.82	218.0	1985.3	27766.1			
5.875	20.42	49.56	233.9	2251.4	33494.1			
6.000	20.90	51.34	250.8	2553.1	40403.5			
6.125	21.39	53.15	268.8	2895.0	48737.8			
6.250	21.88	54.99	288.1	3282.4	58791.0			
6.375	22.37	56.85	308.6	3721.5	70917.3			
6.500	22.85	58.76	330.5	4219.0	85545.1			

TABLE 3.1. ARL as a function of k and h for CUSUM for mean shift of standardized normal data.

k	0.25	0.50	0.75	1.00	1.25	1.50	1.75	2.00
ARL								
125	4.788	3.057	2.179	1.642	1.260	0.950	0.673	0.412
250	5.994	3.716	2.626	1.983	1.550	1.215	0.926	0.660
375	6.732	4.109	2.891	2.182	1.715	1.363	1.067	0.797
500	7.267	4.389	3.080	2.323	1.830	1.466	1.164	0.892
625	7.688	4.608	3.227	2.433	1.919	1.545	1.238	0.963
750	8.034	4.787	3.348	2.523	1.992	1.609	1.298	1.021
875	8.329	4.939	3.450	2.599	2.053	1.662	1.347	1.069
1000	8.585	5.071	3.538	2.665	2.105	1.708	1.390	1.110
1125	8.812	5.187	3.617	2.723	2.152	1.748	1.427	1.146
1250	9.016	5.291	3.686	2.776	2.193	1.784	1.461	1.178
1375	9.201	5.386	3.750	2.823	2.231	1.816	1.490	1.206
1500	9.370	5.472	3.807	2.866	2.265	1.846	1.517	1.232
1625	9.526	5.551	3.861	2.906	2.297	1.873	1.542	1.256
1750	9.670	5.625	3.910	2.943	2.326	1.898	1.565	1.278
1875	9.805	5.693	3.956	2.977	2.353	1.921	1.586	1.298
2000	9.931	5.757	3.999	3.009	2.379	1.942	1.606	1.317

TABLE 3.2. h as a function of k and ARL for CUSUM for mean shift of standardized normal data.

The most immediate use of this table is to assess the ARL implications of different choices of k and h, and as a way of exploring the sensitivity of the CUSUM's in-control behavior to different choices. For example, to design for $k = 0.5$, the choice $h = 4.375$ would be quite good for a target ARL of 500. Increasing h to the rounder figure of 4.5 would increase the in-control ARL to 556, which might be close enough for practical purposes to justify the rounding.

Table 3.2 is a logical inverse to Table 3.1. It lists the h values that will yield various tabulated in-control ARLs, and so is particularly useful in designing CUSUMs. The reference value k of the CUSUM is chosen on the basis of the size of shift that is either anticipated, or that would be large enough to cause serious concern. The in-control ARL is selected on the basis of the tolerable false-alarm rate. The final piece of the puzzle could then be provided by Table 3.2 which converts the (k, ARL) pair to the required h.

Both tables are for one-sided CUSUMs. In other words, they relate to the ARL before an upward CUSUM signal, or equivalently, before a downward CUSUM signal. Most practical problems involve two-sided CUSUMs, in which you maintain an upward CUSUM C_n^+ and a downward CUSUM C_n^-. As we show in Chapter 6, if the two CUSUMs use the same k and h parameters, the ARL until the first of these two CUSUMs signals is half the ARL to signal of either one.

Thus the choice $k = 0.5, h = 4.375$ is listed as giving an ARL of 556. If an upward and a downward CUSUM are maintained, then the ARL until the first of them signals is one half this figure, or 278.

On the design side, if we want the two-sided CUSUM to have an ARL of, say, 500, then we should design so that each of the two component CUSUMs has an ARL double this, or 1,000. This is given, for example, by the choice $k = 0.5, h = 5.071$.

3.2 Out-of-control ARLs

The design of the CUSUM is made by specifying k and the tolerable in-control ARL, from which the h value follows. The key performance measure for a CUSUM is the CUSUM's ability to detect shifts in mean, once they occur. We look first at the most relevant aspect of this — its speed in detecting pure mean shifts — but then show that shifts in variance also have an impact on the chart for the mean. To see both points, we go to broader models for out-of-control behavior.

Suppose that the measurement stream of observations in control is normally distributed with mean μ and standard deviation σ:

$$X_n \sim N(\mu, \sigma^2)$$

In this chapter, we work with individual readings standardized to $N(0,1)$:

$$U_n = \frac{X_n - \mu}{\sigma}.$$

We keep CUSUMs for both upward and downward shifts, but we enrich the notation so that these CUSUMs can have different reference values and decision intervals, giving the schemes

$$\begin{aligned} S_0^+ &= 0 \\ S_0^- &= 0 \\ S_n^+ &= \max(0, S_{n-1}^+ + U_n - k^+) \\ S_n^- &= \min(0, S_{n-1}^- + U_n + k^-), \end{aligned}$$

where k^+ and k^- represent the reference values of the upward and the downward location CUSUMs, respectively. The system of two CUSUMs signals an increase in mean if $S_n^+ > h^+$ and a decrease if $S_n^- < -h^-$

This formulation permits the allowances k^+ and k^- to differ, as well as the decision intervals h^+ and h^-. For CUSUMs of normal data, the upward and downward reference values k^+ and k^- often be the same, as the upward and downward decision intervals h^+ and h^-. This reflects the situation where increases and decreases in mean are of equal concern and should be handled identically. This common situation does not always hold, though.

Suppose X is a "more is better" measure, like the lifetime of a component or the profit on a sale. Then a large decrease in the mean μ reflects an intolerable situation calling for emergency action. An increase in μ would indicate potential for improvement, and lead to a more measured response. Another difference is the likely size of a shift: if the process had been in operation for some time so that there was some experience in running it well, large increases in μ would be unlikely, but many special causes could lead to large decreases in μ. Under these circumstances, it would be reasonable to set k^+ smaller, whereas k^- might be appreciably larger.

Different reference values immediately imply different decision intervals if the upward and downward CUSUMs are to have the same in-control ARL. We may also choose to have a longer in-control ARL for upward shifts than for downward shifts, reflecting that the damage done by poor quality was more urgent than the potential for long-term quality improvement. Even if k^+ and k^- were equal, we might want different values for h^+ and h^-.

3.2.1 Model

Suppose now that the process has a step change in mean and/or variance. At first, we assume that the step change is in effect right from the first observation onward; then we think about the effect of a change at some other instant. Let the distribution after the step change be

$$X_n \sim N(\mu + \Delta\sigma, \lambda^2\sigma^2),$$

where the parameter Δ is used to describe a shift in mean, and λ is used for a shift in variance. If $\Delta = 0$, then the mean is unchanged; if $\lambda = 1$, then the variance is unchanged.

We now show how the out-of-control ARL of the CUSUM following this shift can be calculated using tables or computer programs for the in-control situation. The distribution of the variables U_n used in the standardized CUSUM becomes

$$U_n = \frac{X_n - \mu}{\sigma} \sim N(\Delta, \lambda^2).$$

The upward and downward CUSUMs for the mean have the recursions

$$S_n^+ = \max(0, S_{n-1}^+ + U_n - k^+)$$

$$S_n^- = \min(0, S_{n-1}^- + U_n + k^-).$$

We could think of the U_n variables as linear transformations of some other standard normal quantities Z_n:

$$U_n = \lambda Z_n + \Delta,$$

where $Z_n \sim N(0, 1)$. Writing the recursions in terms of the standard normal quantities Z_n gives

$$S_n^+ = \max(0, S_{n-1}^+ + \lambda Z_n + \Delta - k^+)$$
$$S_n^- = \min(0, S_{n-1}^- + \lambda Z_n + \Delta + k^-).$$

We now construct the following upward and downward CUSUMs of the Z_n,:

$$T_0^+ = T_0^- = 0$$
$$T_n^+ = \max(0, T_{n-1}^+ + Z_n + \frac{\Delta - k^+}{\lambda}) \quad (3.1)$$
$$T_n^- = \max(0, T_{n-1}^- + Z_n + \frac{\Delta + k^+}{\lambda}). \quad (3.2)$$

Comparing the equations for S_n^+ and T_n^+ and those for S_n^- and T_n^- shows that

$$S_n^+ = \lambda T_n^+ \quad (3.3)$$
$$S_n^- = \lambda T_n^- \quad (3.4)$$

and so $S_n^+ > h^+$ if and only if $T_n^+ > \lambda h^+$, and $S_n^- < -h^-$ if and only if $T_n^- < -\lambda h^-$.

What does this mean?

- The ARL of the CUSUM S^+ to the decision interval h^+ is identical to the ARL of the CUSUM T^+ to the decision interval h^+/λ.
- The ARL of the CUSUM S^- to the decision interval h^- is identical to the ARL of the CUSUM T^- to the decision interval h^-/λ.
- The CUSUMs T^+ and T^- are CUSUMs of standard unshifted $N(0,1)$ quantities.
- The CUSUM T_n^+ has reference value $(k^+ - \Delta)/\lambda$ and decision interval. h^+/λ
- The CUSUM T_n^- has reference value $(k^- + \Delta)/\lambda$ and decision interval k^-/λ.
- **The ARL of the CUSUM following a shift therefore does not require any special tables or additional computing ability — it can be done using tables or computer codes for the in-control situation.**

> We can find out-of-control ARLs using in-control tables or software for a modified reference value and decision interval.

3.2.2 The ARL following a shift in mean

We start out with the most important situation, that in which the mean changes but the variance does not. So Δ is nonzero, but λ equals 1. Then the out–of–control ARL of the upward CUSUM with reference value k and decision interval h equals the in–control ARL of a CUSUM with reference value $k - \Delta$ and decision interval h.

Example

Suppose we are using the CUSUM with $k = 1$ and $h = 2.375$. This CUSUM is "tuned" to detect a 2 standard deviation shift in the mean. Its in–control ARL is given by Table 3.1 as 556, so this would be a reasonable design choice if the target in-control ARL were 500 and an approximate design using just Table 3.1 (rather than software or the more directly useful Table 3.2) was adequate.

> The ARL following a shift in mean can then be read from the same row of Table 3.1 by moving to the column headed by the value of $k - \Delta$.

If the mean shifts upward by $\Delta = 0.25$ standard deviations, then the ARL can be read from the table as the in-control ARL of a CUSUM with $k = 1 - 0.25 = 0.75$ and $h = 2.375$. The ARL of this CUSUM is 169.7. Continuing leftward in this row gives the ARL following other shifts:

Shift (sd)	0	0.25	0.50	0.75	1.00	1.25
ARL	555.5	169.7	59.3	24.7	12.5	7.54

The shift Δ is measured in standard deviations. Thus shifts in the original measurements of X need to be converted to multiples of σ before this table lookup can be done.

So far we have not emphasized the *sign* of Δ. The previous example considered Δ values larger than zero, and it is indeed upward shifts in the mean that the upward CUSUM is designed to detect. It is possible for the upward S_n^+ CUSUM to cross its decision interval even if Δ is negative. The ARL to this event can likewise be looked up in Table 3.1 by moving right rather than left from the pivot entry. Using this same row of the table, the ARL until a signal on the S_n^+ chart is:

Shift (sd)	-0.25	-0.50
ARL	1982	7414.

These ARLs are large, as one would certainly hope. The CUSUM S_n^+ for an upward shift in mean will, if left long enough, cross its decision interval, *even if the mean has shifted downward.* Fortunately, the ARL until this mistake is made is very large.

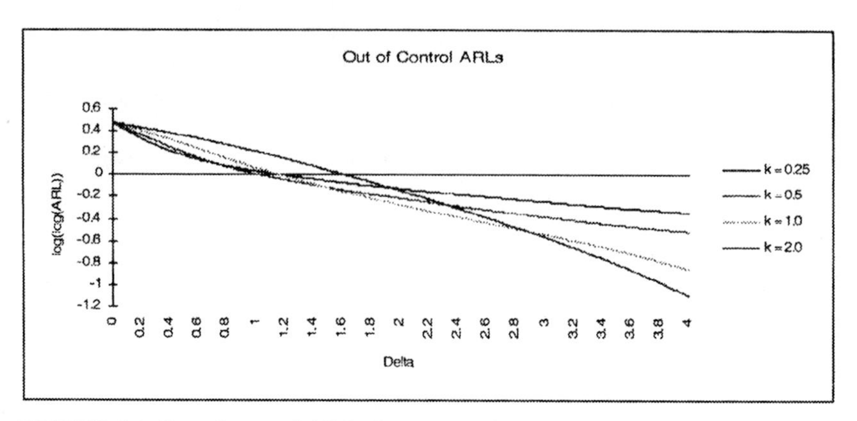

FIGURE 3.1. Out of control ARLs for various k choices. The vertical scale is the logarithm (base 10) of the logarithm of the ARL. The horizontal scale is Δ, the size of the actual mean shift.

3.2.3 *ARL sensitivity to choice of k*

Unlike the Shewhart chart, the CUSUM is tuned for maximum sensitivity at a particular selected shift. This additional strength is an advantage that includes an implied drawback: it raises the question of how much performance is lost if the process mean shifts to some level other than that for which it was tuned. Some feeling for this question can be found in Figure 3.1. This involves CUSUMs with an in-control ARL of 1,000, with k values of 0.25, 0.5, 1, and 2. The figure shows the ARL of each of these four CUSUMs in response to a range of shifts Δ from 0 up to 4.

The vertical scale of Figure 3.1 is unusual. It shows the $\log_{10}$ of the $\log_{10}$ of the ARL, a scale that was chosen to more-or-less linearize the ARL profiles and to make it easier to see where each of the CUSUMs performs well and where poorly.

Figure 3.1 shows that each of the CUSUMs has its own "place in the sun" — a range of nonzero Δ values for which its ARL is smaller than those of the other CUSUMs. Reading the ARLs from the original table form rather than from the graph of Figure 3.1 shows that the CUSUM with $k = 0.25$ is the best of the four for all Δ less than 0.73. The $k = 0.50$ CUSUM, which is tuned to shifts of size $\Delta = 1$ then takes the lead and is the best of the four in the range $0.73 < \Delta < 1.46$. The CUSUM $k = 1$ leads in the range $1.46 < \Delta < 2.87$, and the $k = 2$ CUSUM is the best of these four for Δ values above 2.87.

These ranges in which each of the four CUSUMs is the best are intuitively in line with their being the optimal diagnostics for shifts of size $\Delta = 0.5$, 1.0, 2.0, and 4.0 respectively. It is of some possible interest that the crossover values of Δ at which each CUSUM surpasses its neighbor are not exactly midway between these values to which the CUSUMs are tuned, but lie slightly to the left of the midway value.

3.2.4 Out-of-control states and two-sided CUSUMs

Until now, we have concentrated on detecting an upward shift in mean using the S^+ upward CUSUM, but showed how the ARL of the upward CUSUM following a downward shift in mean could be computed.

It is clear from symmetry that the response of the downward CUSUM S^- to downward shifts in mean is exactly the same as that of the equivalent upward CUSUM S^+ to an equal upward shift in mean. Thus, for example, we saw that the upward CUSUM S^+ with $k^+ = 1$ and $h = 2.375$ had an ARL of 59.3 following an upward shift of $\Delta = 0.5$ in mean. The symmetry of the normal distribution and the equivalence of the definition of the S^+ and S^- CUSUMs show that the downward CUSUM S^- with reference value $k^- = 1$ and decision interval $h^- = 2.375$ have an ARL of 59.3 before signaling following a one half standard deviation downward shift in mean.

In most applications, we want to monitor changes in mean in either direction. So we keep two decision interval CUSUMs: S_n^+ for increases and S_n^- for decreases. The pair of CUSUMs are then run until one of them crosses its decision interval, at which point the process is stopped and examined for special causes. What is the ARL of this combined system that stops when the first CUSUM signals?

We write ARL^+ for the ARL of the upward CUSUM, and ARL^- for the ARL of the downward CUSUM, and ARL for the ARL of the combined system. Starting as we do with $S_0^+ = S_0^- = 0$, Van Dobben de Bruyn (1968) showed that

$$\frac{1}{ARL} = \frac{1}{ARL^+} + \frac{1}{ARL^-}. \tag{3.5}$$

(This generalizes the in-control result that we mentioned earlier.) To find the overall ARL of the upward/downward pair would then require that we include the contribution from the "opposite" side as well as that on the "correct" side. Return, for example, to the CUSUM $k = 1, h = 2.375$. If we use the same chart parameters for the upward and downward CUSUM, then as we noted earlier, the in-control ARL of the pair is $556/2 = 278$.

Now suppose that the process mean changes by $\Delta = 0.25$ standard deviations. We know $ARL^+ = 169.7$ and $ARL^- = 1982$ from the example earlier in the section. Then the ARL of the combined scheme is given by

$$\frac{1}{ARL} = \frac{1}{ARL^+} + \frac{1}{ARL^-} = \frac{1}{169.7} + \frac{1}{1982}$$

so that the overall scheme has an ARL of 156. This is some 10% smaller than the 169.6 for the S^+ scheme; the reduction comes about because of the rather remote possibility that the real upward shift be marked by a downward signal on the CUSUM.

> We can use Equation 3.5 to find the ARL of the combined charts for both in–control and out–of–control ARLs.

3.3 FIR CUSUMs: zero start and steady state start

3.3.1 *Introduction*

In all the discussion so far, the CUSUM was started at zero: $S_0 = 0$. Lucas and Crosier (1982) suggested a way to improve the performance of the CUSUM for the special situation in which the process mean is already shifted at the time the CUSUM charting begins. This is to give the CUSUM a "head start."

The "head start" or "Fast Initial Response (FIR)" CUSUM for upward shifts is defined by

$$\begin{aligned} S_0^+ &= H^+ \\ S_n^+ &= \max(0, S_{n-1}^+ + X_n - k^+) \end{aligned}$$

and signals if $S_n^+ > h^+$.

The quantity $H^+ > 0$ is the "head start." (Setting $H^+ = 0$ gives us the same zero-start CUSUM we have been using until now.) The intuitive motivation for the FIR CUSUM is that, if the process is already out of control when the charting begins, then starting the CUSUM part way toward the decision interval will hasten the signal. If, however, the process is not out of control, then it is likely that the CUSUM soon drop back to zero, after which the FIR CUSUM behaves like a conventional zero-start CUSUM.

Following some experimentation with different head starts, Lucas and Crosier recommended that the head start H^+ be one-half the decision interval h^+. This choice $H^+ = h^+/2$ has become generally accepted as the standard.

On the "no free lunch" principle, of course, we have to pay for this improved initial response somewhere; the price is that to maintain the same in-control ARL it is necessary to increase the decision interval h somewhat. Table 3.3 is a FIR equivalent to Table 3.2, and shows the revised h values necessary for the FIR CUSUM to attain each target in-control ARL.

For example, with $k = 2$ and in-control ARL of 1,000, the zero-start CUSUM has a decision interval $h = 1.110$, whereas the FIR CUSUM has an almost identical decision interval $h = 1.111$.

Over on the other side of the table, at $k = 0.25$ and the same in-control ARL of 1,000, the zero-start CUSUM used $h = 8.585$ while the FIR CUSUM uses $h = 8.709$ — a much larger difference, although it is not at

k	0.25	0.50	0.75	1.00	1.25	1.50	1.75	2.00
ARL								
125	5.001	3.133	2.216	1.663	1.272	0.957	0.677	0.415
250	6.180	3.777	2.656	2.000	1.559	1.221	0.930	0.662
375	6.900	4.162	2.916	2.196	1.723	1.369	1.071	0.799
500	7.422	4.437	3.103	2.336	1.838	1.471	1.167	0.893
625	7.832	4.652	3.248	2.445	1.926	1.549	1.241	0.965
750	8.170	4.828	3.367	2.534	1.998	1.613	1.300	1.022
875	8.458	4.977	3.468	2.609	2.059	1.666	1.350	1.070
1000	8.709	5.107	3.556	2.675	2.111	1.712	1.392	1.111
1125	8.931	5.222	3.633	2.733	2.157	1.752	1.429	1.147
1250	9.130	5.324	3.702	2.785	2.199	1.787	1.463	1.179
1375	9.311	5.417	3.765	2.831	2.236	1.820	1.492	1.208
1500	9.477	5.502	3.822	2.874	2.270	1.849	1.519	1.233
1625	9.630	5.581	3.874	2.914	2.302	1.876	1.544	1.257

TABLE 3.3. h as a function of k and ARL for FIR CUSUMs for the mean of a standardized normal observation. Here the head start is given by $H = h/2$.

once evident whether this larger h corresponds to any substantial loss of performance.

To address this performance question, we need to think about the out-of-control ARL of the FIR CUSUM.

> FIR results in much faster detection if the process starts out of control at the cost of slightly slower detection if it starts in control.

3.3.2 *Out-of-control ARL of the FIR CUSUM*

Going back to the general framework, suppose the process measure X_n has shifted in mean and/or standard deviation so that the standardized quantities U_n are $N(\Delta, \lambda^2\sigma^2)$.

The upward CUSUM with head start H^+ has the recursion:

$$\begin{aligned} S_0^+ &= H^+ \\ S_n^+ &= \max(0, S_{n-1}^+ + U_n - k^+). \end{aligned}$$

We set up the CUSUM of the standardized quantities Z_n as

$$\begin{aligned} T_0^+ &= \frac{H^+}{\lambda} \\ T_n^+ &= \max\left(0, T_{n-1}^+ + Z_n - \frac{k^+ - \Delta}{\lambda}\right). \end{aligned}$$

By inspection, $S_n^+ = \lambda T_n^+$ for all n, and so the ARL until S_n^+ crosses a boundary h^+ is the same as the ARL until the T_n^+ CUSUM crosses the boundary h^+/λ.

So we find once again:

- The ARL of the out-of-control process measures U_n can be calculated using in-control results.
- The ARL of the S^+ CUSUM is the in-control ARL of a CUSUM of $N(0,1)$ quantities with reference value $(k^+ - \Delta)/\sigma$, decision interval h^+/λ, and head start H^+/λ.
- If there is no scale shift ($\lambda = 1$), then the out-of-control ARL of the CUSUM is found by just shifting the reference value by Δ.

A FIR CUSUM for a downward shift is defined in exactly the same way:

$$
\begin{aligned}
S_0^- &= -H^- \\
S_n^- &= \min(0, S_{n-1}^+ + X_n + k^-)
\end{aligned}
$$

and signals if $S_n^- < -h^-$.

By symmetry, this CUSUM has exactly the same properties as does the upward CUSUM with the same parameters

The usual value of the head start H^+ is half the decision interval h^+. This same proportionality holds between the head start and the decision interval of the CUSUM of Z_n used to find the out-of-control ARL of the FIR CUSUM of X_n. The GETARL procedure from the Web site provides in-control ARLs of FIR CUSUMs with this default head start $H = h/2$. It can therefore be used to handle both the in-control and out-of-control performance.

Example

We return to the FIR CUSUM with $k = 0.25$ and $h = 8.709$ and in-control ARL 1000. Suppose the process mean shifts immediately after startup. The performance of the FIR CUSUM for shifts of 0.25 and 0.5 standard deviations is given by this output from GETARL

```
error code 0 k=  .0000  h= 8.7090  regular ARL  97.50 FIR ARL  73.37
error code 0 k= -.2500  h= 8.7090  regular ARL  31.57 FIR ARL  19.39
```

Looking first at the rightmost column, a shift of $\Delta = 0.25$ standard deviations occurring immediately the CUSUM starts up has an ARL of 73.37 to detection. A half standard deviation shift has an ARL to detection of 19.39.

The other value returned by GETARL is a zero-start ARL. If the process is in control at startup, then once the first excursion above the axis ends,

S_n^+ revert to zero. If a mean shift of 0.25 standard deviations happens at some stage when $S_n^+ = 0$, then the ARL to detection be 97.5. Similarly, a shift of 0.5σ occurring when the CUSUM is in the zero state have an ARL to detection of 31.57.

Comparing these two figures, the FIR feature certainly does seem to enhance the CUSUM's ability to pick up mean shifts that are already in place when the CUSUM starts.

The price to be paid for this performance improvement is in the larger h. If the mean shift does not occur immediately when the CUSUM starts out but at some later time, then the benefit of the head start be lost — but the crossing boundary h is permanently raised. We can explore the effect of this higher h using GETARL to check zero-start CUSUMs using the zero-start h. For this example, this gives:

```
error code  0 k=  .0000  h= 8.5850 regular ARL  95.07 FIR ARL  71.55
error code  0 k= -.2500  h= 8.5850 regular ARL  31.08 FIR ARL  19.12
```

For a 0.25 standard deviation shift, the ARL of the zero-start CUSUM from either the initial start or from any later time when the CUSUM happens to be at zero is 95.07. If the shift happens when the FIR CUSUM is back at zero, then the ARL to detection is 97.50, a larger figure but only marginally. For a half-standard deviation shift, there is even less of a difference. If the FIR CUSUM has gone back to zero at the time of the shift, the ARL is 31.57, whereas for the zero-start CUSUM it is 31.08.

The situation of a shift when the CUSUM first starts out is obviously the most favorable for the performance of the FIR scheme relative to the zero-start CUSUM. Here its performance is substantially better than that of the zero-start CUSUM. Conversely, a particularly unfavorable situation for the FIR scheme against the zero-start one is when the first excursion has come to an end and the CUSUM is back at zero when the mean shift occurs. Here, the edge in favor of the zero-start CUSUM is very slim.

> Using a head start routinely is a good idea. It can produce big improvements in CUSUM performance when the process mean shifts before the CUSUM reverts to zero, but costs only a minimal penalty if the process mean occurs later.

3.3.3 ARL of two-sided FIR CUSUMS

Suppose (as is usually the case) that we maintain a pair of CUSUMs for upward and downward changes. The recursions are

$$S_0^+ = H^+$$

$$\begin{aligned} S_0^- &= -H^- \\ S_n^+ &= \max(0, S_{n-1}^+ + U_n - k^+) \\ S_n^- &= \min(0, S_{n-1}^- + U_n + k^-) \end{aligned}$$

with a signal if $S_n^+ > h^+$ or $S_n^- < -h^-$.

The ARL of this combined scheme is discussed in very general terms in Yashchin (1985). Writing $A^+(s)$ for the ARL of the upward CUSUM with a head start of s and $A^-(s)$ for the ARL of the downward CUSUM with head start $-s$, (in both cases $s = 0$ is the zero-start ARL), then under suitable conditions, the ARL of the combined scheme is

$$ARL = \frac{A^+(H^+)A^-(0) + A^-(H^-)A^+(0) - A^+(0)A^-(0)}{A^+(0) + A^-(0)}.$$

The condition required for this result is that

$$k^+ + k^- \geq \max(H^+ + H^- - \min(h^+, h^-), |h^+ - h^-|).$$

This obscure-looking condition is needed to ensure that the upward and downward CUSUMs do not "interact." If the condition is met, then if the upward CUSUM signals, the downward CUSUM be at zero, whereas if the downward CUSUM signals the upward CUSUM be at zero.

The condition is met automatically in the situation we have been concentrating on, where $H = h/2$ for both upward and downward CUSUMs, and $k^+ = k^-$ and $h^+ = h^-$. The condition becomes less self-evidently true when we set up different CUSUMs for upward and downward shifts in normal mean, and also when we go to some other distributions where the upward and downward CUSUMs routinely have different parameters, and in both these situations it is necessary to check that the condition holds.

Example

As an example, consider the CUSUM $k = 0.25, h = 8.709$. Supplying these parameters to GETARL gives

```
error code 0 k= .2500  h= 8.7090 regular ARL  1066.44 FIR ARL  1000.04
```

If we design both the upward and downward CUSUMs with this value of the parameter pair, then the combined CUSUM have ARL

$$ARL = \frac{2 \times 1000 \times 1066 - 1066^2}{2 \times 1066}$$

which simplifies to 467. Notice that this is a bit smaller than half the ARL of either side. When the upward and downward CUSUMs have the same ARL, then the general expression for the combined scheme's ARL simplifies to

$$ARL = A^+(H^+) - A^+(0)/2$$

which would give this same result with less calculation and, perhaps, more insight; the combined scheme's ARL is less than half the FIR scheme's ARL by half the difference between the FIR and zero-start ARL.

3.3.4 Initial and steady-state ARL

The effect of the head start raises another issue that we just hinted at in Chapter 2. We have discussed the ARL from when the CUSUM starts to when it crosses the decision interval. In the case of zero-start CUSUMs, this is also the ARL from any future instant at which the CUSUM happens to have returned to zero.

There are other ways to think about ARL. Consider starting the run from some arbitrary starting time, at which the CUSUM may or may not be at zero, until the signal. Of particular interest is this "steady state" ARL of the CUSUM. Initially, the CUSUM starts at its starting value (zero or a head start value) and then changes in response to the variations in the process reading stream. This means that the value of the CUSUM S_n^+ is a random variable. Its distribution depend on n, but as n increases the impact of the starting value of the CUSUM wanes, and S_n^+ approaches some statistical distribution — the "steady state distribution" that does not depend on n. If S_n^+ exceeds h along the way, then we assume that the CUSUM is restarted at zero without interruption.

The steady-state distribution is an unusual one. There is a nonzero probability that $S^+ = 0$ (and this probability increases with k and can be close to 1 for large k values), and nonzero S^+ follow some continuous distribution on the interval $(0, h)$.

Although not having any standard easily recognized form, this steady-state distribution of S^+ is available. It is a by-product of the Markov chain approach set out in a later chapter for computing the ARL.

The steady-state ARL of the CUSUM is the ARL measured from a random starting point, where the starting point follows this steady-state distribution. We can usefully think about the steady-state ARL as the ARL of a FIR CUSUM whose head start, rather than being a constant, is a random variable following the steady-state distribution of S^+. This conceptual method also provides a computational algorithm for calculating the steady-state ARL by "mixing" the distribution of FIR ARLs over this distribution of the head start.

3.4 Controlling for the mean within a range

We have discussed the CUSUM as a diagnostic for checking whether the mean moves away from its in-control level of μ by any nonzero amount. In some problems, we do not need or want so sharp a detection tool. For example, if the specification of a bolt implies that bolts are perfectly acceptable as long as their mean diameter lies between, say, 0.495 and 0.505 mm, then we do not care if the mean changes from one level to another — provided both levels are within this range.

This allows us to incorporate our concerns about the process specifications into the construction of our control scheme.

We can formulate this problem as

$$X_n \sim N(\mu, \sigma^2)$$

with the process being regarded as in control (in terms of the specifications) provided μ lies between some lower limit μ_L and some upper limit μ_U.

The decision interval CUSUM is easily adapted to handle this situation. The design consists of deciding how far the mean must increase above μ_U before it becomes problematic. Write μ_N for this unacceptable level.

For clarity in exposition we work in the original scale of X rather than the usual standardized scale. The upward CUSUM is then designed using the reference value and scheme:

$$\begin{aligned} k^+ &= \frac{\mu_U + \mu_N}{2} \\ C_n^+ &= \max(0, C_{n-1}^+ + X_n - k^+). \end{aligned}$$

As usual, the reference value is chosen midway between the in-control value and the out-of-control value for which the CUSUM should have maximum sensitivity. This recipe differs from what has gone before only in that the value selected for the in-control mean is the largest such acceptable mean.

Similarly, for decreases in mean we set a reference value k^- midway between μ_L and some lower mean sufficiently far below μ_L to cause concern. Then the downward CUSUM is defined by

$$C_n^- = \min(0, C_{n-1}^- + X_n - k^-).$$

We do not specify the starting values for the CUSUM; they can be either zero start or head start. Those who prefer the V-mask form of the CUSUM to the zero-start decision interval format can use it in this problem also. The CUSUM is defined by

$$\begin{aligned} C_0 &= 0 \\ C_n &= C_{n-1} + X_n - \frac{\mu_L + \mu_U}{2}. \end{aligned}$$

The fact that we only want to control μ to within a range does not affect the construction of the chart, but is taken into account in the construction of the V-mask that may be used to diagnose it. The slope of the legs of the V-mask is increased by $(\mu_U - \mu_L)/2$ beyond what would be used to control the mean to a single in-control level.

3.4.1 Example

One of the concerns in a chemical laboratory is that the assays (chemical readings) in a particular measurement run may be wrong because of some problem with the setup of the instrument, the reagents used, or a procedural error. One common check on problems of this sort is to periodically run a "standard." This consists of taking some material from a large fixed pool and analyzing it as an unknown. If a particular run is unable to produce an acceptable value for the standard, then it is concluded that run had some error and it is repeated. There are two sorts of problems to diagnose. A procedural error in an individual run would be indicated by that particular run's standard being unusually high or low. A smaller but persistent shift in the mean of the standard would indicate that there was some bias affecting all assays. You therefore want to check the values of the standard both for isolated special causes, and for persistent shifts in mean.

The ability of the CUSUM to detect small persistent shifts can become something of a liability in this problem. Some small drifts in mean are inevitable, and it is counterproductive to try to keep the mean constant to the degree that a CUSUM cannot detect any shift at all. A sensible approach then is to set up the CUSUM to permit the mean to drift within some acceptable range and only start to signal when this range is left.

A data set provided by Dr. Daniel Levine, of the Rogosin Institute, illustrates this problem. The series consists of a triglyceride standard. Figure 3.2 is a time-ordered plot of 149 values. Visually, the sequence seems to show some outliers (isolated bad runs) and to rise at the end of the sequence.

We set up the problem of monitoring the mean to see if it is within the band (115,120)mg/dL. We use the in-control standard deviation 3mg/dL, and set the CUSUM up for detection of a drift of more than 1 standard deviation outside this range: that is, to above 123mg/dL or below 112mg/dL. This then gives the reference value $(120 + 123)/2 = 121.5$ for the upward CUSUM, and $(115 + 112)/2 = 113.5$ for the downward CUSUM.

Table 3.4 shows the calculation of the first few points of the decision interval CUSUM, and Figure 3.3 gives the full CUSUM. The CUSUM shows several interesting features. First, to a much greater extent than in the CUSUMs seen previously, there is very little action in the early part of the chart. Since the mean appears to be well within its allowable range, setting the reference value by adding an allowance to the edges of the range has ensured that most of the DI CUSUM values are zero. The CUSUM does show occasional spikes caused by the "outliers", individual assays

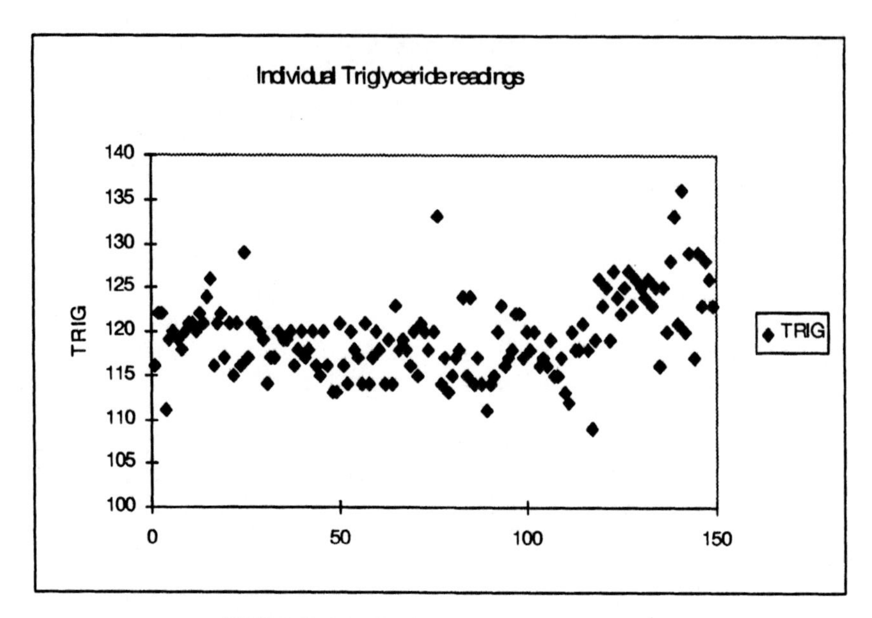

FIGURE 3.2. Triglyceride standard values.

Run	Assay	C^+	C^-
1	116	0.0	-0.0
2	122	0.5	-0.0
3	122	1.0	-0.0
4	111	0.0	-2.5
5	119	0.0	-0.0
6	120	0.0	-0.0
7	119	0.0	-0.0
8	118	0.0	-0.0
9	120	0.0	-0.0
10	121	0.0	-0.0
11	121	0.0	-0.0
12	120	0.0	-0.0
13	122	0.5	-0.0
14	121	0.0	-0.0
15	124	2.5	-0.0
16	126	7.0	-0.0

TABLE 3.4. Some calculations for the triglyceride example.

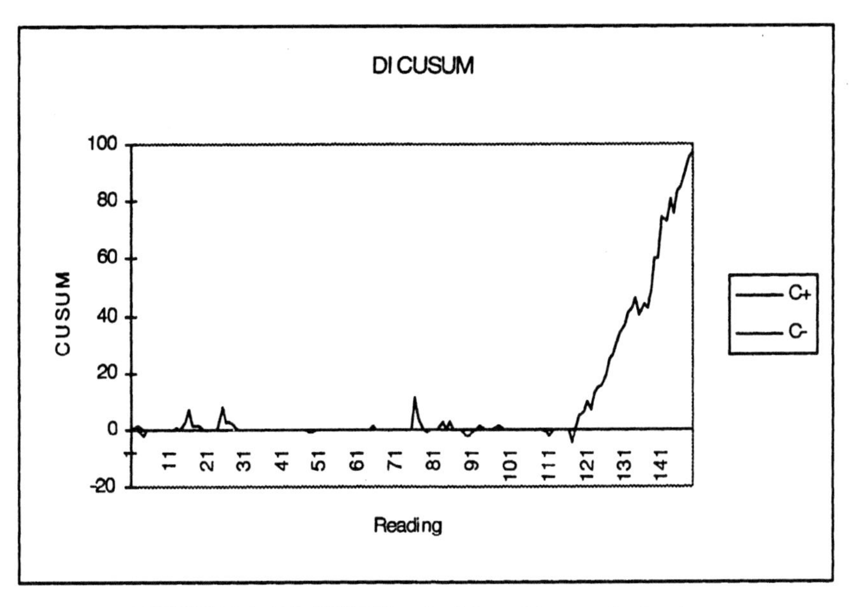

FIGURE 3.3. DI CUSUMs of triglyceride standard values.

whose values were outside the range you would expect for a reading whose mean lay between 115 and 120 and whose standard deviation was 3. These outliers are the individual bad runs which, in standard laboratory practice, would be repeated. The final and most important feature of the CUSUM is the sharp rise of the C^+ CUSUM toward the end of the series. The CUSUM shows that the mean broke out of the range (115,120) convincingly somewhere around run number 120.

The slope of the C^+ CUSUM is about 3mg/dL, which, with the reference value of $k^+ = 4$, shows that the mean triglyceride assay appears to have increased about the time of run 120 to $117.5 + 3 + 4 = 124.5$mg/dL.

> The within-range CUSUM has been successful in both signaling the increase in mean soon after it occurred, and in staying well within its decision interval limits during the earlier part of the sequence when the mean was within tolerance.

3.5 The impact of variance shifts

Except for the introductory derivation, in all the detailed discussions so far we have assumed that the change from the in-control to the out-of-control situation took the form of a change in mean.

Let's consider briefly what happens if the mean is unchanged, but the variance of the data changes, specifically

$$X_n \sim N(\mu, \lambda^2\sigma^2).$$

As shown in the general discussion, the S^+ CUSUM following this change is the CUSUM of standard normal quantities, but using reference value k/λ and decision interval h/λ instead of k and h.

It may be initially surprising that this change has a substantial impact on the ARL. We can illustrate this using GETARL to calculate some ARLs based on the $k = 0.5, h = 5.071$ CUSUM. Modifying the reference value and decision interval by the scale factor λ gives:

```
lambda = 1.0   k=   .5000  h=  5.0710 ARL is    1000.30
         1.1   k=   .4550  h=  4.6100 ARL is     445.21
         1.2   k=   .4170  h=  4.2260 ARL is     241.00
         1.3   k=   .3850  h=  3.9010 ARL is     149.72
         1.4   k=   .3570  h=  3.6220 ARL is     102.04
         1.5   k=   .3330  h=  3.3810 ARL is      74.75
```

If the standard deviation increases by 50%, the ARL drops from 1,000 to 75. As some checks with GETARL show, this is the same ARL you would get if the mean increased by 0.35 standard deviations.

> An increase in variance is a major cause for concern. Quite apart from the importance of a variance change in its own right, an increase in variance dramatically increases the chances of a false alarm on the CUSUM for location.

This gives two potent arguments for using CUSUMs to check for changes in the variance of the data alongside those checking for changes in the mean.

On first hearing about the sensitivity of the CUSUM for mean to changes in variance, many people assume this is a problem unique to the CUSUM. Not so — the Shewhart chart is equally (and maybe even more) susceptible to misinterpreting a variance change as a shift in mean. In a Shewhart Xbar chart for a normal process, an increase in the underlying standard deviation of 50% has the same effect as bringing in the control limits from 3 standard errors to $3/1.5 = 2$ standard errors. This brings the ARL for an upward signal from 740 to 40.

As Shewhart said, all quality control charts for mean need to be supplemented with control charts for variance. The CUSUM chart is no exception. Not only are variance changes important in their own right, but also unnoticed increases in variance can lead to wrong signals of changes in mean. We sketch two approaches to controlling variance shifts in the context of CUSUMs.

> Variance shifts have dramatic effects on both Shewhart and CUSUM charts for the mean.

3.5.1 Individual data — the approximate normal transform

CUSUMs for normal mean are generally best made on individual observations rather than rational group means. This raises the question of how to check for departures from the in-control variance. We outline one approach now, and a later chapter provides a more powerful methodology within a broader framework.

If the data X_n are $N(\mu, \sigma^2)$, then the standardization

$$U_n = \frac{X_n - \mu}{\sigma}$$

produces a stream of $N(0,1)$ variables. Hawkins (1981) found that the transform $\sqrt{|U_n|}$ is very close to normally distributed with mean 0.822 and variance 0.119; in other words, the variable

$$W_n = \frac{\sqrt{|U_n|} - 0.822}{0.349}$$

is very close to $N(0,1)$. If the *variance* of X_n were to increase, however, then the *mean* of W_n would increase, and this increase could be detected by a CUSUM of the W_n. This idea leads to the upward and downward CUSUMs for scale

$$\begin{aligned} V_0^+ &= V_0^- = 0 \\ V_n^+ &= \max(0, V_{n-1}^+ + W_n - k^+) \\ V_n^- &= \min(0, V_{n-1}^- + W_n + k^-) \end{aligned}$$

(which could also be started with head-starts rather than zero starts). The CUSUMs S^+ and S^- of U_n, and V^+ and V^- of W_n provide CUSUM analogs of the Shewhart Xbar and R chart, a pair of charts that monitor location and scale simultaneously.

Perhaps the greatest virtue of this proposal is that the location and scale CUSUMs can be plotted together on the same chart, since the summands

n	X_n	U_n	W_n	S_n^+	S_n^-	V_n^+	V_n^-
1	46.68	-0.66	-0.02	0.00	-0.16	0.00	0.00
2	50.02	0.00	-2.17	0.00	0.00	0.00	-1.67
3	54.58	0.92	0.39	0.42	0.00	0.00	-0.79
4	49.64	-0.07	-1.59	0.00	0.00	0.00	-1.87
5	38.18	-2.36	2.05	0.00	-1.86	1.55	0.00
6	45.21	-0.96	0.45	0.00	-2.32	1.50	0.00
7	51.11	0.22	-1.01	0.00	-1.60	0.00	-0.51
8	51.45	0.29	-0.81	0.00	-0.81	0.00	-0.82
9	49.71	-0.06	-1.67	0.00	-0.37	0.00	-1.98
10	60.05	2.01	1.71	1.51	0.00	1.21	0.00

TABLE 3.5. Calculation of the combined location and scale CUSUM

U_n and W_n are both $N(0,1)$ while the process is in control (the W only approximately), giving a single chart that shows the control of both mean and variance. Of course to do this, the charts need to use the same reference value and decision interval, but choosing k and h on the basis of what is best for controlling μ, and then just going along with the same values for the scale chart gains much simplicity and generally loses little performance.

A second advantage of this proposal is that it greatly simplifies the ARL calculations. Our software routines to find ARLs (which work well for the χ^2 distribution with three or more degrees of freedom) do not work well for the one and two degree of freedom cases.

The calculations for this combined location/scale chart can be automated using the DOCUSUM program. Some data illustrate the calculations of the combined location and scale CUSUM. The original data stream X_n are normal with in-control mean 50 and standard deviation 5, but after randomly sampling the first 50 readings from this distribution, we increased the variance for the rest of the series while leaving the mean at 50. The first 10 data values, the corresponding U_n and W_n, and decision interval CUSUMs with $k = 0.5$ (on both upward and downward CUSUMs) are shown in Table 3.5. Figure 3.4 shows the combined CUSUMs of the whole sequence.

Figure 3.4 shows quite clearly how the variance increased around the middle of the sequence. Another feature that can be seen is an illustration of how the scale change can affect the location CUSUM. The location CUSUM went down beyond the middle of the sequence and crossed the decision interval for the ARL=1000 CUSUM. That this signal was due to the scale shift rather than a true location shift is indicated by the fact that the location CUSUM soon returned to the axis, while the scale CUSUM continued to increase.

This method of scale control has some advantages: the ability to plot location and scale CUSUMs on the same chart with the same decision interval is the most obvious. It also has some deficiencies. It is not fully

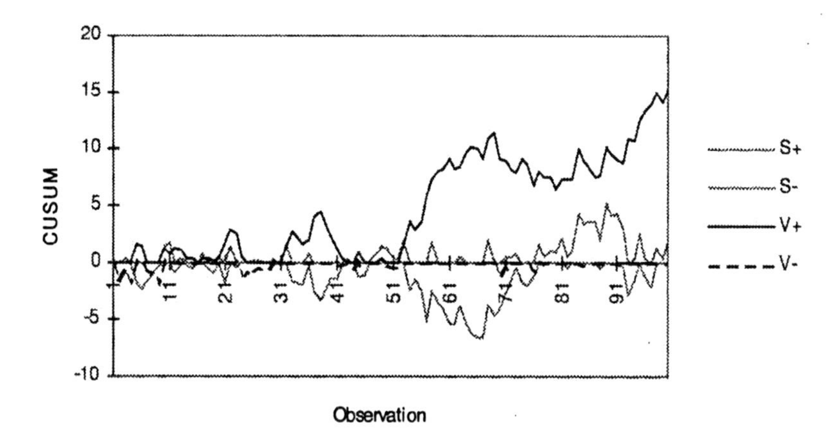

FIGURE 3.4. Combined location and scale CUSUMs for normal individual observations example.

statistically efficient; it "throws away" about 30% of the information about σ in the data. It is also somewhat affected by location shifts. If the mean of the data shifts, then the mean of the W_n will also move away from zero, although by a much smaller amount. This means that after a large shift in mean, the scale CUSUM may give false signals. This is a relatively minor difficulty however, given that the signal on the scale CUSUM is likely to be accompanied by a much larger one on the location CUSUM.

Finally, this CUSUM is quite effective for detecting *increases* in variance, but it is not nearly so effective in detecting *decreases*; in fact, its ARL following small decreases in variance can actually exceed its in-control ARL. In practical terms, this means that although it is quite well suited to signaling process problems (increased random variability), it is not as good for detecting process improvements (reduced random variability). Since uncontrolled variability is the major reason for most quality problems, variance reduction is the way to go for process improvement, and the fact that the transformation to normality does not detect improvements very well certainly should count against it.

3.5.2 Rational groups — variance CUSUMs

For the purpose of monitoring the process mean, rational groups are in general better than individual readings only when there is a substantial economy of scale in taking several process readings at the same time. Still, if we do take rational groups of size greater than 1, it becomes possible to make CUSUM charts of the variances of the rational groups. Suppose that at time n we sample a rational group of size m, getting process readings X_{nj}, $j = 1, \ldots, m$. Reduce each group to its mean $\overline{X}_n$ and sum of squared deviations

$$D_n = \sum_{j=1}^{m} (X_{nj} - \overline{X}_n)^2.$$

Then D_n/σ^2 follows a χ^2 distribution with $m-1$ degrees of freedom.

A scale CUSUM R can then be defined by

$$\begin{aligned} R_0^+ &= H^+ \\ R_0^- &= -H^- \\ R_n^+ &= \max(0, R_{n-1}^+ + D_n - k^+) \\ R_n^- &= \min(0, R_{n-1}^- + D_n + k^-) \end{aligned}$$

with a signal of upward movement in variance if $R_n^+ > h^+$, and one of downward movement in variance if $R_n^- < -h^-$.

We spend much of Chapter 4 on CUSUMs of χ^2-distributed data, so we spend no more time here on this CUSUM. It is enough for now to mention that this CUSUM is the optimal diagnostic for detecting a step shift in the variance of normal data if we assume nothing about their mean, so from the viewpoint of performance it is ideal.

There is a variant that can also be considered: this is to use the sum of squared deviation from the in-control *true mean* μ, forming a CUSUM of the quantity

$$D_n^* = \sum_{j=1}^{m} (X_{nj} - \mu)^2$$

which, by the analysis of variance identity, can be written

$$D_n^* = D_n + m(\overline{X}_n - \mu)^2.$$

The distribution of D_n^*/σ^2 is χ^2 with m degrees of freedom (rather than the $m-1$ of D_n. Its performance in terms of detecting variance shifts is therefore the same as what we would get using a rational group of size $m+1$ with the cumulative summand D_n, effectively increasing the size of the rational group, apparently without having to pay for it.

Of course there is a price to be paid for the improved performance. If the mean shifts from μ to some other level, then the distribution of D_n is unaffected, and so the scale CUSUM continue to monitor scale changes without any harmful side-effects of the location shift. This is not true of

D_n^*, whose distribution go from a central to a non-central χ^2, and which therefore tend to drift upward, increasing the likelihood of a false alarm on the scale CUSUM. Perhaps for this reason, CUSUMs of D_n^* are not common, with the scale CUSUM more usually being of D_n.

The location CUSUM S is affected by scale changes, as noted earlier; a variance increase can dramatically reduce the ARL of the location CUSUM even in the absence of a shift in mean. This potential confusion between the two is resolved by checking the scale CUSUM when there has been a signal on the location CUSUM, since the variance increase show up sooner and more sharply on the scale CUSUM than on the location CUSUM. The same is true of CUSUMs using D^*. If there is a step change in mean, it tend to show up much more strongly on the location CUSUM than on the scale CUSUM, so that the potential for confusion is quite small. You may well question then why users should be unwilling to use a scale CUSUM that has some shadows of location effects, while living with the fact that all charts for location are affected by scale shifts.

> CUSUM charts for variance are mandatory for effective process control, and are discussed in detail in Chapter 4.

3.6 Combined Shewhart and CUSUM charts

This monograph is focused on CUSUM charts. CUSUMs are excellent diagnostics for detecting and diagnosing step changes in process parameters. These are not the only changes that can occur, however. Transient special causes are also an important reality, and an important source of quality problems. They cannot be ignored, and relying solely on CUSUM charts for SPC is shortsighted. Just as Shewhart charts are not particularly effective in detecting less-than-massive persistent changes in process parameters, so CUSUM charts are not particularly effective in detecting less-than-massive transient changes in process parameters.

Proper SPC requires the use of both types of control: CUSUMs for persistent changes, and Shewhart charts for transient problems. These charts can and should be made using the same process measurement stream, as this provides the most effective use of the information gathered and helps ensure that diagnoses made using the CUSUM and the Shewhart charts are mutually consistent.

When you use process measurements for combined Shewhart and CUSUM control, then it is necessary to think again about the issue of rational groups. Suppose one reading is taken every two hours on a process working three eight-hour shifts a day. For purposes of CUSUM charting, it makes little difference whether these four readings per shift are aggregated into a

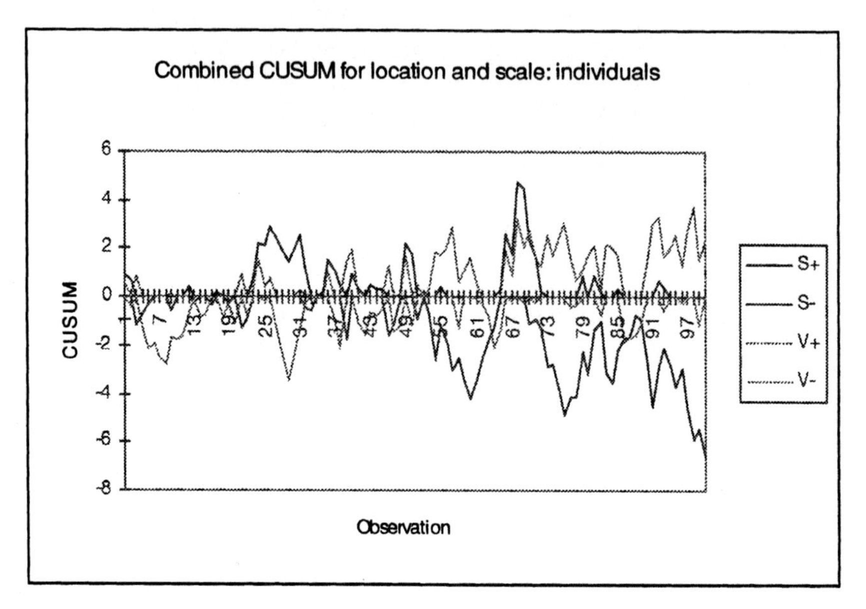

FIGURE 3.5. CUSUMs for location and scale of individual readings.

single point for the location chart or are accumulated individually, but this is not the case for the Shewhart chart. Here you need to aggregate readings to get acceptable performance. The basic idea of rational grouping — of putting together all the readings and only the readings that are likely to be affected by a special cause — leads you to analyze the four readings from each shift as a rational group. The mean and the standard deviation (or the range) of the four readings then be plotted on Shewhart Xbar and S (or R) charts as single points representing each shift.

The CUSUM and Shewhart charts should be scaled so that their time scales line up, and they should be kept together so that any signal on any of the charts can be interpreted in the light of information from the other charts.

3.6.1 Example

The following example illustrates diagnoses from simultaneous CUSUM and Shewhart charts. The sequence of 100 readings can be regarded as either 25 rational groups of size 4, or as 100 individual readings. Figure 3.5 shows the CUSUM of data regarded as individual readings, as they would normally be analyzed. Figures 3.6 and 3.7 show the Xbar and S charts of the same data, but regarded as 25 rational groups of size 4.

A decision interval of $h = 4.77$ would give an in-control ARL of 740 with the choice of $k = 0.5$ used for the CUSUM; this matches the in-control ARL of the Shewhart charts. Figure 3.5 clearly shows a decrease in mean

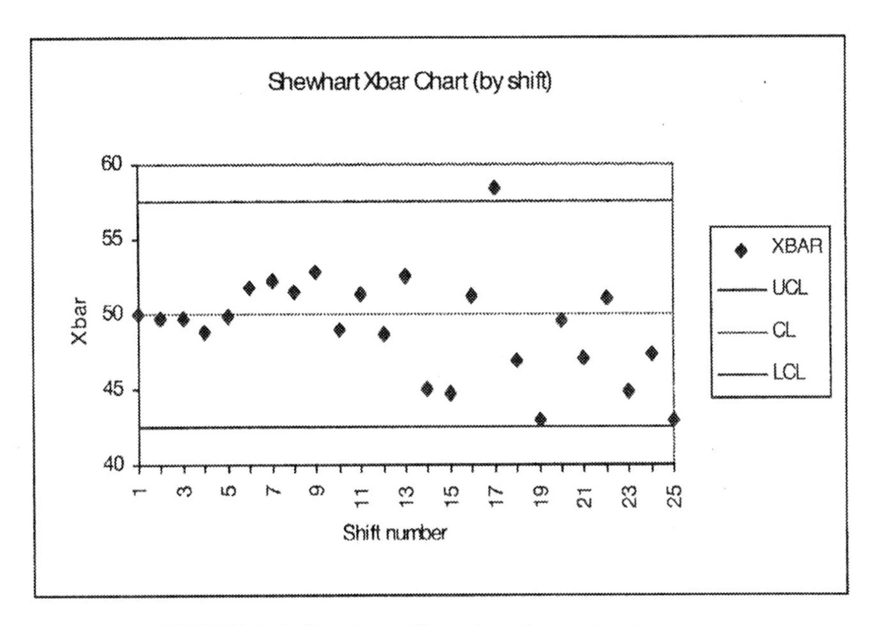

FIGURE 3.6. Shewhart Xbar chart for rational groups.

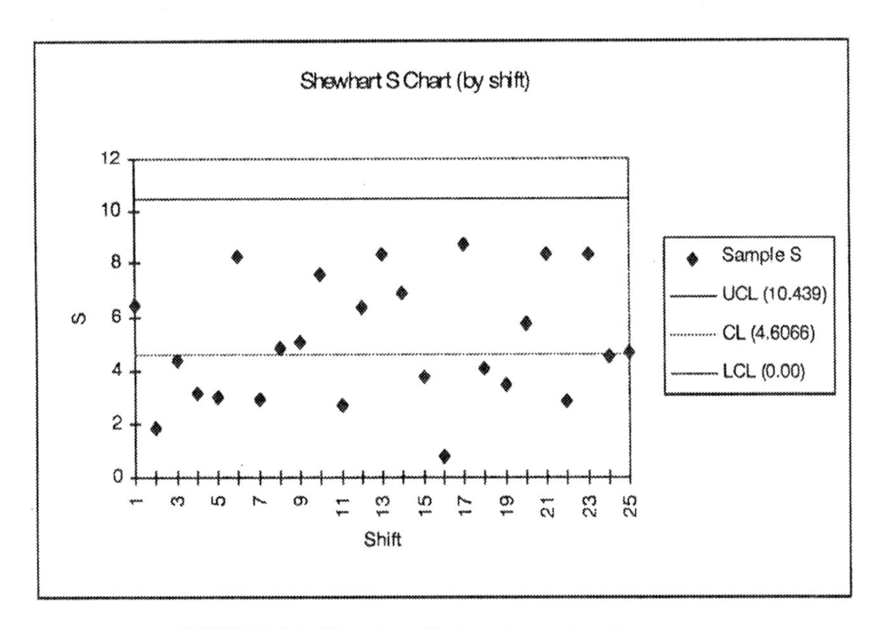

FIGURE 3.7. Shewhart S chart for rational groups.

toward the end of the sequence. The decision interval is breached around the 95th observation, with the diagnosis that the mean moved downward just before the 70th observation. This is the only CUSUM signal for change in either location or scale.

The Xbar chart picks up a quite different feature. The point for shift 17 goes beyond the upper control limit. This diagnoses an individual shift in which the process readings were high, apparently right before the mean shifted downward.

Although the data are fictional, the situation shown is not. Some problem occurred during shift 17, but the real cause was not found. In an attempt to correct the problem, the process was adjusted in such a way that the mean went down. This effect is too small to show up on the Shewhart chart, but is detected by the CUSUM chart.

> This example illustrates the dangers of tinkering with a process without full understanding of the actual special causes operating on it.

3.7 Effect of model departures

We examine some common model departures and their effects on the CUSUM procedures we have outlined.

3.7.1 Nonnormality

One of the concerns when charting data is the dependence of the ARL on the statistical distribution of the readings. Modeling of process data has historically not been a high priority in the SPC community, as there is a feeling that the behavior of control charts is not influenced all that much by the actual distribution of the data. There is some truth to this, but it would be an overstatement to say that statistical distributions are irrelevant. If a CUSUM scheme is designed for an in-control ARL of 1,000 and the actual ARL is 500, then there be double the number of false alarms anticipated, and this would be troubling. If the actual ARL were 1,200, then there would likely be no cause for concern. It does seem valuable to know which of these possibilities to expect.

To shed some light on this, consider three CUSUMs designed to have an in-control ARL of 1,000 with normal data. They are $k = 1, h = 2.67, k = 0.5, h = 5.07$, and $k = 0.25, h = 8.19$. Using the CUSARL code from the Web site, we then calculate the true ARL for various data distributions:

- a heavy-tailed symmetric distribution: the Laplace distribution with density function $f(x) = 0.5\exp(-|x|)$;

	Example $k = 1$	$k = 0.5$	$k = 0.25$
Laplace	241	585	933
χ_2^2	82	195	435
χ_9^2	163	357	624
Uniform	24789	1435	1069

TABLE 3.6. ARL for various distributions using normal-based design with nominal ARL of 1000.

- a heavy-tailed skew distribution: the χ^2 with 2 degrees of freedom;
- a somewhat less skew distribution: the χ^2 with 9 degrees of freedom;
- a symmetric light-tailed distribution: the uniform over the range (0,1).

The true ARLs are shown in Table 3.6. This table shows that the actual ARLs have quite a lot of structure. First, there is cause for concern in that the true ARL can indeed be a long way from the design value of 1,000. Tail weight is obviously important. All three heavy-tailed distributions give ARLs that are below the design value, whereas the light-tailed uniform distribution gives ARLs above the design value. The CUSUM parameters have a huge impact on the ARL. Going from left to right, for each distribution the ARLs move steadily toward the design value. This suggests a general principle that with small k values and large h the impact of nonnormality is much less severe than it is with larger k. The basic reason for this behavior is not so much the small k but the large h which makes it impossible for one or two values from far out into the tail of the distribution to trip the decision interval. This being so, signals tend to have to come from a number of observations, and the central limit theorem tames the tail behavior.

The table gives cause for some concern about the effect of using normal-based k and h values when the data in fact follow some substantially nonnormal distribution. The effect of the nonnormality is greatest when the CUSUM is tuned to detect large shifts (which also suggests that nonnormality affect Shewhart charts even more than it does these CUSUMs), but even with CUSUMs tuned for small shifts the in-control ARL can be badly off target if the data distribution has tail weight markedly different than the normal.

There is another more technical, theoretical concern about using normal-based chart designs for nonnormal data, and this is the statistical issue of efficiency. As the general theory of Chapter 6 shows, a CUSUM of the "right" function of the data gives the optimal test for a step change in the distribution. Among the many things that are inferior to this best test is a CUSUM of the "wrong" function of the data.

We mentioned this briefly in the discussion of controlling a normal variance. The optimal CUSUM for scale in normal data is a CUSUM of the variances of the rational groups. Even though a transformation of the sample variance may give a quantity that is close to normally distributed, a CUSUM of this normal transform is not as powerful as a CUSUM of the variances themselves. Using a less effective diagnostic is like deliberately throwing away some of the process readings, an obviously wasteful practice. More careful modeling of the statistical distribution of process readings translates into better use of the information they contain and is equivalent to achieving the same performance as you could with more data but without the need for the additional data. It is hard to find *any* good argument against doing this.

> CUSUMs designed for normal processes can have a much higher false alarm rate for skewed or heavy-tailed data.

3.7.2 Independence

The CUSUM also rests on the assumption that the process readings are statistically independent of one another. This raises the questions of what effect correlation has on the ARLs, and how one might better control those series of readings that do have serial correlation.

We look at some numbers dealing with the first of these issues. The simplest model for nonindependent data is a Markov chain, in which observations spaced r time units apart have a correlation of ρ^r, where ρ is the autocorrelation of lag 1. This autocorrelation may take any value between -1 and 1. Positive values (the most common in industrial processes) indicate that the process tends to drift slowly, staying above its mean for several observations at a time, and staying below for several observations. Negative values of ρ arise in processes that overcorrect, with an above-average reading tending to be followed by one below average.

Finding the ARL of a CUSUM of autocorrelated data using general-purpose ARL software is not as easy as the other calculations we have done so far, but it is straightforward to do using simulation. We did this for a range of ρ values, getting the ARL for the same three schemes with their nominal in-control ARL of 1,000 that we used to investigate nonnormality. The results of this are shown in Table 3.7

This table is also quite sobering. An autocorrelation of, say, 0.2 is not very large, yet it is enough to reduce the ARL from its nominal 1,000 to below 400.

A common piece of advice when analyzing correlated data is to choose sampling times sufficiently far apart that there be no appreciable correlation between the successive values. We do not much like this advice. If process considerations suggest that sampling at a two-hour interval is advisable, then increasing the sampling interval to, say, eight hours on the

ρ	$k=1$	$k=0.5$	$k=0.25$
-.50	8452	1408451	1960784
-.40	7148	146843	76278
-.30	5391	24149	17328
-.20	3250	6309	5333
-.10	1807	2201	2048
.00	992.8	995.4	999.6
.10	597.8	517.4	557.5
.20	385.0	309.8	348.5
.30	259.7	201.0	233.9
.40	185.8	140.1	165.6
.50	140.0	103.5	123.2

TABLE 3.7. ARL by simulation for various values of ρ using normal-based design with nominal ARL of 1,000. We see negative autocorrelation increases ARL; positive autocorrelation decreases ARL.

grounds that it simplifies the statistics seems stupid. A much sounder approach is to work on making the statistics fit the real problem and not vice versa.

We sketch some ideas on CUSUMs of autocorrelated data in a later section.

3.8 Weighted CUSUMs

In some problems, each process reading X_n is paired with a "weight" W_n. A familiar example is a record of a vehicle's fuel consumption, where the amount of fuel put in the tank on refill n, X_n, is recorded along with the distances traveled W_n since the last refill.

The statistical model underlying the weighted CUSUM is

$$X_n \sim N(W_n\mu, W_n\sigma^2).$$

If W_n were integer-valued, then we could think of X_n as being a sum of W_n independent $N(\mu, \sigma^2)$ variables. The actual weights W_n are not necessarily integer-valued, but we may still find this "partial sum" idea helpful on thinking about how to handle the X_n and their associated W_n.

The V-mask form of the CUSUM is more easily motivated than the decision interval form. It defines the CUSUM

$$C_n = \sum_{j=1}^{n}(X_j - W_j\mu)$$

and plots C_n against the running total $T_n = \sum_1^n W_j$.

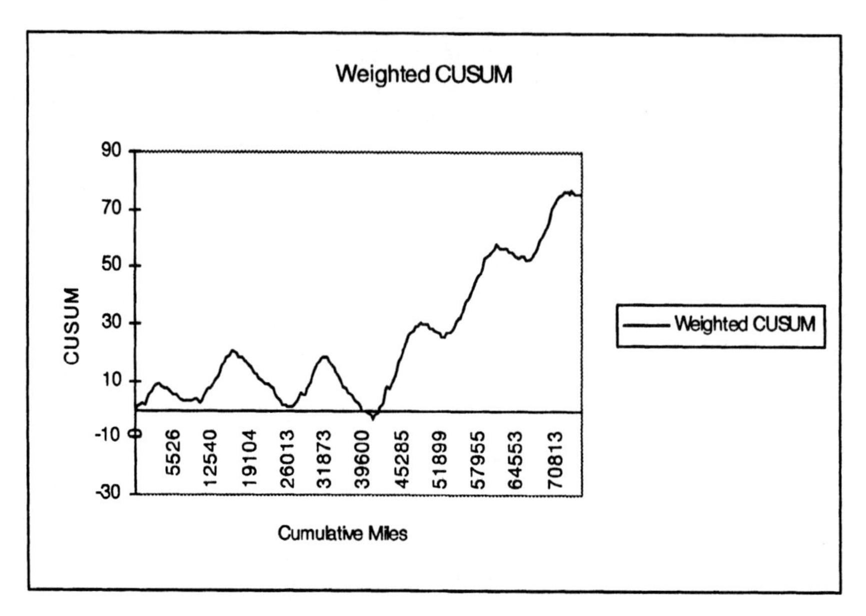

FIGURE 3.8. Weighted CUSUM of fuel consumption data.

It is easy to see that C_n has expectation zero and variance $T_n\sigma^2$.

As an example of a V-mask weighted CUSUM, we consider some fuel consumption data gathered by Michele Ireland over a period of some 6 years for her automobile. The data consist of the number of gallons of gasoline X_n put into the tank at the nth refueling, and the number of miles traveled W_n since the last fillup.

The first 100 tanks, covering 41,271 miles, yield an average fuel consumption of 0.0197 gallons per mile; we use this as an estimate of μ. We can also estimate σ^2 as

$$\hat{\sigma}^2 = \frac{\sum W_j(X_j - W_j\hat{\mu})^2}{\sum W_j}.$$

This turns out to be 137.82/41271, giving an estimate of σ of 0.0578 gallons per mile. We assume that this model is correct, and that these estimates are indeed the true values of μ and σ.

The V-mask CUSUM of the entire data set is given in Figure 3.8. This shows two major structural features in the data. One is the sharp upward segment starting shortly after the 40,000 mile mark. Apparently the fuel consumption increased at this stage. As the CUSUM looks to center on a sloping straight line apart from the waves it has over its whole length, it would appear that the fuel consumption underwent a step change. A quick eye-estimate would be that the last point on the CUSUM is (75,000; 75), and since the CUSUM passed through (40;000; 0), this gives a slope of 0.0021 gallons per mile for the second segment of the CUSUM. Inverting the before and after figures, the average fuel consumption for the first 40,000 miles was 50.8 miles per gallon, and that for the subsequent 35,000 miles averaged 45.9 miles per gallon, a roughly 10% increase in consumption.

The other striking feature of the CUSUM is the waves. These are tied in with the seasonal changes — the fuel consumption is higher during the winter months than it is during summer. It would be possible to use the CUSUM to estimate these seasonal differences too. For example the CUSUM seems to have a seasonal peak around (17,000; 20) before dropping to (26,000; 0), a slope of -0.0022 gallons per mile. The seasonal difference therefore looks to be about 10% also.

The slow waves in this CUSUM show that the model of independent normal variables with constant mean is not very plausible for this data set, and this, along with the clear change in fuel consumption in the latter half of the series, makes it somewhat academic to discuss estimation of the variance σ^2. In other problems, though, we would be more interested in formal testing for a shift in level, and this leads us to the decision interval form of the CUSUM.

The decision interval form of the CUSUM is useful for formal testing in situations (unlike that illustrated by this fuel consumption data set) where the model of independent identically distributed in-control data is reasonable. We define the DI CUSUM in the original units of the problem, rather than scaled units. The DI CUSUM is to be designed for an upward shift in the mean of X_n from $W_n\ \mu$ to $W_n\ (\mu+\delta)$. Set the reference value, as usual, halfway between the in-control and the out-of-control means; that is, to

$$k_n = W_n(\mu + (\delta/2)).$$

Define the upward CUSUM by

$$\begin{aligned} C_0^+ &= 0 \\ C_n^+ &= \max(0, C_{n-1}^+ + X_n - k_n). \end{aligned}$$

The CUSUM signal if $C_n^+ > h^+$, where h^+ is some decision interval.

It is hard to set h to fix the in-control ARL in completely general terms. In fact, the whole situation starts out with a conceptual difficulty of how we measure the ARL. It could be in the number of points on the CUSUM that are added before the signal; it could equally well be in terms of the total value of W_n that accumulates before the signal. In the fuel consumption context, we could measure the interval between signals in the number of times the fuel tank is refilled (the first option) or the total miles traveled (the second option). Neither of these choices is necessarily the right one. For either choice, getting an approximate h is quite easy, although the accuracy of the approximation depends on how variable the W_n are. Write $\overline{W}$ for the "average" W_n (strictly speaking, the expected value of W_n, assuming that they follow a common independent statistical distribution, but more realistically a reasonable common value). Then for design purposes, we approximate the distribution of X_n by $N(\overline{W}\mu, \overline{W}\sigma^2)$.

Approximating the out-of-control mean by $\overline{W} \cdot (\mu + \Delta\sigma)$, the reference value k_n is then offset from the in-control mean by $\overline{W}\Delta\sigma$ units, which corresponds to

$$k = \frac{\overline{W}\Delta\sigma}{2\sqrt{\overline{W}}\sigma} = \frac{1}{2}\sqrt{\overline{W}}\Delta$$

$$(\Delta = \delta/\sigma)$$

standard errors of the summand. From this k, software or tables can be used to get the required h for a target in-control ARL. This ARL be measured in units of points on the CUSUM, not of accumulated W. We then use the decision interval

$$h^* = h\sqrt{\overline{W}}\sigma.$$

3.8.1 Example

Ignoring the issue of the seasonal effects, let us go through this exercise for the gas consumption data. We tune the CUSUM for a shift in fuel consumption from the calibration mean of 50.8 miles per gallon to 45 miles per gallon, this corresponds to a shift in μ from 0.0197 to 0.0222, a change of $\delta = 0.0005$ gallons per mile.

We estimated σ at 0.0571 giving $\Delta = 0.0087$. The mean of W in the calibration data is 413 miles — we use this for $\overline{W}$, giving finally $k = \sqrt{413} \times 0.0087 = 0.177$.

We could design for an in-control ARL measured in tanks of fuel, or we could base the ARL on total miles. For example, 50 tanks is about 18 months of driving, and might be a good choice for the in-control ARL. Or we might select 20,000 miles of driving for the in-control ARL, a figure that corresponds to 20,000/413 = 48 tanks. Whichever approach we choose, we can convert between tanks and miles as the units of ARL using $\overline{W}$.

The ANYGETH design code gives the output

```
Enter k and ARL values (zeroes stops run)
.177 50
k=  .177,  h= 3.8555  ARL=     50.02
```

so that $h = 3.86$ produce an acceptable in-control ARL. Converting this back to the scale of the weighted CUSUM gives $h^* = h\sigma\overline{W} = 3.854 \times 0.0571 \times 20.3 = 4.47$.

The decision interval weighted CUSUM is shown in Figure 3.9. Because of the seasonal effects the decision interval is breached as every winter approaches, but in addition to this seasonal ebb and flow of the fuel economy,

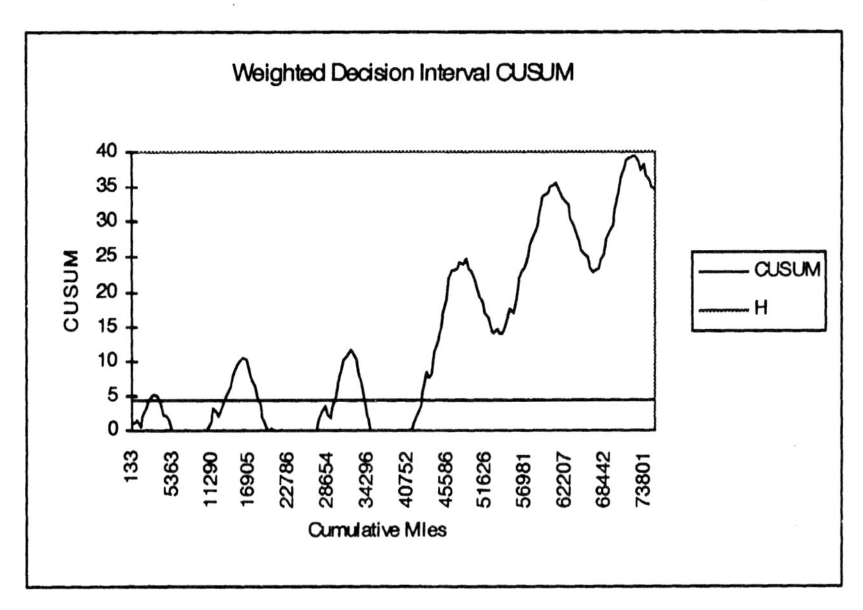

FIGURE 3.9. Weighted decision interval CUSUM for gas consumption example.

the CUSUM clearly shows the large increase in fuel consumption starting after the 40,000 mile mark.

3.9 Summary

This chapter extended the development of the CUSUM for normal data. We found methods to determine out-of-control ARLs, methods to improve the responsiveness of the CUSUM, and methods for controlling the mean within a range of allowable values. We analyzed the effect of model departures, including variance shifts, nonnormality, and autocorrelated data. We examined the advantages of combining CUSUM and Shewhart charting, and finally closed with a discussion of weighted CUSUMs.

CUSUMs represent a fairly mature technology, and the normal distribution, as the primary distribution for quality control, has been well studied. There are other occasions when other distributions better model the data, and we have argued for the use of accurate models earlier.

In the next chapter, we turn to CUSUMs for many of these other distributions, beginning with two continuous distributions.

3.10 Further reading

Kemp (1967) illustrated the effect of nonnormality of readings on their CUSUM. The effect of normality on the performance of control charts is a source of much disagreement. Shewhart chart practitioners often allude to the "control limit theorem", a folk theorem to the effect that the Shewhart Xbar chart is robust to departures from normality. The discussion earlier in this chapter cast doubt on this, showing by example how nonnormality of readings could affect the ARL of the CUSUM quite significantly, and that CUSUMs with large k were more vulnerable than those with small k. As the Shewhart chart is equivalent to a CUSUM with a large k, this suggests that it is more sensitive to the normality assumption than the typical CUSUM.

The weighted normal CUSUM was defined by Bissell(1973) in a slightly more general context than we are using here. In addition to our model (where the variance in Y_n is proportional to W_n), he set out a "components of variance" model with an additional variance unrelated to W_n. Yashchin's (1989) weighted CUSUMs include one in which the weights are associated with recency rather than precision, a choice leading to a CUSUM that is effective in detecting linear rather than step changes in mean.

We suggested that when you have individual readings you control for scale changes using the differences $(X_n - \mu)$, CUSUMming either their squares or the square root of their absolute value. This has the drawback of making the scale control susceptible to contamination from shifts in the mean. Another possibility is using the moving range, as is done with Shewhart charts of individual readings. The moving range is defined by $|X_n - X_{n-1}|$. CUSUMs of this quantity, or of its square, can be used for a scale control that is unaffected by mean shifts; the down sides of this approach are that it loses some statistical efficiency, and that the successive moving ranges are correlated.

4

Other continuous distributions

> Many years ago I called the Laplace-Gaussian curve the NORMAL curve, which name, while it avoids an international question of priority, has the disadvantage of leading people to believe that all other distributions of frequency are in one sense or another "abnormal." That belief is, of course, not justifiable.
>
> *Karl Pearson*

Until this chapter, we have looked almost exclusively at CUSUM schemes for the normal distribution. We now take a detailed look at two other continuous members of the exponential family of distributions (the gamma and the inverse Gaussian) and their CUSUMs. We find that the design of CUSUMs for these distributions differs from the normal distribution; in particular,

$$k \neq \frac{\mu_1 + \mu_2}{2}$$

and sometimes the optimal CUSUM involves CUSUMming a transformed variable.

The (rather technical) theory for these complications is deferred until Chapter 6. In this chapter, we focus on the "how" and "when," saving the "why" for a later optional chapter.

Why these two distributions? The gamma family is important as a model in its own right, but more especially as the distribution that describes the sample variance of a sample from a normal population. The inverse

Gaussian distribution is a second distribution useful for modeling positive, right-skew data produced by a variety of processes.

4.1 The gamma family and normal variances

In Section 6.2.2, the CUSUM schemes for the gamma distribution are derived. The CUSUM scheme for normal variances is given as a special case in the same section. In this section, we give some background on CUSUMs of gamma distributed variates, discuss a few points of design, and present examples. Our focus here is to use the results of Section 6.2.2 without the added burden of the derivation of the results.

4.1.1 Background

When do industrial processes follow the gamma distribution? Several scenarios come to mind.

First, the exponential distribution is a special case of the gamma distribution, and many service times, failure times, and interarrival times are well modeled by the exponential distribution. We know, as well, that the sum of independent identically distributed exponential variables has a gamma distribution.

Second, the gamma distribution is often a plausible model for positive skew data on its own merits.

Third, the chi-squared distribution is also a special case of the gamma distribution. Since the sample variance of a sample from a normal population is known to be proportional to a chi-square random variable, this important case is also studied as a special case of the gamma distribution.

We saw in Chapter 3 the effects of using normal CUSUM schemes on skew and heavy–tailed data. This underscored that the appropriate selection of a statistical model for the process is important for obtaining reliable results.

We have chosen to parameterize the gamma distribution with the two parameters α and β. α is a shape parameter; β is a scale parameter. Say, $Y \sim \Gamma(\alpha, \beta)$. Then the expected value of Y is $E(Y) = \alpha\beta$, and its variance is $VAR(Y) = \alpha\beta^2$.

We note that there are other parameterizations of the gamma distribution in the literature. One characterization of our parameterization is that the parameter of the exponential distribution is the mean, not the inverse of the mean, for the process. A second characterization of our choice is that increases in either α or β result in increases of the mean $E(Y)$.

Recall that sums of gamma-distributed random variables with common scale parameter are also distributed as gamma random variables. Let $Y_i \sim$

$\Gamma(\alpha_i, \beta)$, with the Y_i independent. Then we have:

$$\sum_{i=1}^{n} Y_i \sim \Gamma\left(\sum_{i=1}^{n} \alpha_i, \beta\right) \tag{4.1}$$

We may apply Equation 4.1 in several useful ways. First, let $Y_i \sim \text{Exp}(\beta) \sim \Gamma(1, \beta)$, the exponential distribution with mean β. Then the sum of m independent identically distributed exponential random variables has the distribution $\Gamma(m, \beta)$, also known as the m–Erlang distribution. The sample average of the m observations has the distribution $\overline{Y} \sim \Gamma(m, \beta/m)$.

Either characterization of the sample, by sum or average, allows us to construct CUSUM charts for small persistent changes in β for the exponential distribution based on a gamma distribution CUSUM scheme for changes in β.

A chi-squared random variable with n degrees of freedom is a member of the gamma family:

$$\chi^2_n \sim \Gamma(n/2, 2).$$

The sum of independent chi-squared random variables is also a chi-squared random variable, by application of Equation 4.1.

A scaled chi-squared distribution is also a member of the gamma family, and a possible choice for modeling positive skew data. The scaled chi-squared distribution is also used to express prior and posterior beliefs about the variance in Bayesian analysis of normal processes.

For all these reasons, CUSUM schemes for the gamma distribution are useful tools for the quality professional. The most important application, however, is the CUSUM for the variance of a normal process, to which we now turn.

4.1.2 Normal variances

Consider an industrial process with an output characteristic that is of interest. We assume for this discussion that the output characteristic is well modeled by the normal distribution:

$$Y \sim N(\mu, \sigma^2).$$

We discussed in Chapter 3 the impact of variance shifts on the CUSUM for location. The ARL for the CUSUM for location was shown to be significantly shortened by modest shifts in the variance. For example, we saw that a shift in the variance by a scale factor of 1.5 resulted in a decrease in ARL from 1,000 to 74.75 for the standard scheme with $k = 0.5$ and $h = 5.071$.

Variance shifts are significant in their own right for economic reasons. Consider a process that provides bulk product to a consumer in a container. A 10 pound bag of sugar comes to mind. For reasons of integrity, regulatory penalty, and customer satisfaction, one wishes to specify that the probability of underfilling a container below some threshold be acceptably low. To accomplish this, one sets the mean of the process to be a number of standard deviations above the threshold. For a 10 pound bag, the threshold would be 10 pounds, with the average weight of product for the bags being somewhat higher. This essentially overfills most of the bags to reduce the probability of underfilling any one bag.

An increase in process variability for a fixed average increases the probability of an underfill. To compensate, one must increase the average for the bags, increasing the expected costs of overfilling. A decrease in variability offers the opportunity to reduce the average overfill and still meet the same quality standard, reducing the cost of overfills. Both changes are of interest, although since the costs of underfilling seem to be higher, we are usually more interested in that type of model departure.

This overfill example generalizes to many processes. Increased variability increases costs and usually decreases quality. Reduced variability increases quality and offers the opportunity for reduced costs. We want to know quickly when either situation arises, so we can take the appropriate actions to remedy the situation and to increase our process knowledge.

Increased variability increases costs and reduces quality.

4.1.3 Design of the CUSUM for scale

In Chapter 3, we deferred detailed study of the CUSUM for scale for normal processes until this chapter. We turn first to its design.

As with all our CUSUM schemes, we assume that we have accurately estimated our process parameters. We have that

$$Y \sim N(\mu, \sigma^2).$$

We draw a rational subgroup of size n, and for now we have $n > 1$. Compute the sample mean and variance of this rational group; these summary numbers $\left(\overline{Y}, S^2\right)$ are the "sufficient statistics" of the rational group, and distill all the information about mean and spread in the data. Our optimal CUSUM for a change in σ^2 is given in Equation 6.19. We define X_n as the sample variance of the nth sample, $X_n = S_n^2$. We emphasize: this is a CUSUM of the sample variances, not the sample standard deviations.

To monitor for an increase in variance from the in-control level σ_0^2 to a larger variance σ_1^2, we use the CUSUM

$$\begin{aligned} C_0^+ &= H^+ \\ C_n^+ &= \max(0, C_{n-1} + X_n - k^+) \end{aligned} \tag{4.2}$$

$$k^+ = -\frac{2\ln(\sigma_0/\sigma_1)\sigma_0^2\sigma_1^2}{\sigma_0^2 - \sigma_1^2}. \tag{4.3}$$

This CUSUM has an optional head start H^+. Just as in the case of normal means, we may use a head start to improve the response to shifts that happen soon after the start of the data collection, or we may set H^+ to zero and use the more traditional zero-start CUSUM. The CUSUM for a downward shift in variance ($\sigma_1 < \sigma_0$) has very much the same form:

$$C_0^- = -H^-$$

$$C_n^- = \min(0, S_{n-1} + X_n - k^-) \tag{4.4}$$

$$k^- = \frac{2\ln(\sigma_0/\sigma_1)\sigma_0^2\sigma_1^2}{\sigma_0^2 - \sigma_1^2}. \tag{4.5}$$

We saw that the normal distribution lent itself to standardization. If $Y_n \sim N(\mu, \sigma^2)$, then we could transform to the standardized quantity $U_n = (Y_n - \mu)/\sigma$, which follows a $N(0,1)$ distribution. A CUSUM of U_n is equivalent to one of the Y_n, and has some attractions: it, its reference value, and its decision interval are dimensionless. The gamma distribution also has a standard form, and this can be used to standardize the CUSUM into dimensionless units. The standardization is to use

$$V_n = X_n/\sigma_0^2 = S_n^2/\sigma_0^2$$

whose distribution is that of a chi-squared random variable with degrees of freedom $m - 1$, divided by its degrees of freedom.

This leads to the standardized CUSUM

$$S_0^+ = H^+$$

$$S_n^+ = \max(0, S_{n-1} + V_n - k^+)$$

$$k^+ = \frac{2\ln(\sigma_0/\sigma_1)\sigma_1^2}{\sigma_0^2 - \sigma_1^2}.$$

The C CUSUM and its decision interval can be found by scaling up the corresponding quantities for the S CUSUM by the factor σ_0^2. As with the normal mean situation, this standardization is particularly convenient for tables relating k, h, and the ARL, since these depend only on m in the standard form and not on σ_0.

As before, we select the out-of-control values of the variance for which we desire maximum sensitivity. Those values completely determine our reference value k when substituted into Equation 4.3 or 4.5.

The Web software is able to handle this CUSUM also. They can be used for design purposes (to get the needed h for a target ARL) and for exploring the effects of different h choices.

We illustrate variance CUSUM design with a now-familiar example.

4.1.4 Example: Sugar bags

We return to our notional process that fills sugar bags. In control, the fill for each bag is normally distributed with a standard deviation of $\sigma_0 = .05$ pounds. We want to set the mean of the process so that only 1 in 1,000 bags has less than 10 pounds of sugar. We then want to have maximum sensitivity to detect a shift in the variability of the process that results in the chance of underfill being reduced to 1 in 100. In control, we want our ARL before a false upward signal to be 500.

What should be our design of the CUSUM? How long does it take to detect a shift to the out-of-control state once it occurs?

We go through the calculations for the design of the CUSUM that satisfies these criteria, and determine the ARL in and out of control using our software.

We set the mean of the process to be $\mu = 10.15$, a level reached by noting that the value 3.09 on the standard normal curve cuts off a probability of 1 in 1,000 so that the mean fill weight must be 3.09 standard deviations above the label value of 10 pounds. This works out to a mean fill weight of $10 + 3.09(.05) = 10.15$ pounds.

An increase in the variance increases the proportion of bags that are underweight. We design the CUSUM for maximum sensitivity to a variance that makes the underweight proportion equal to 1 in 100. We complete the following chain of calculations to determine that the corresponding standard deviation is $\sigma_1 = .0789$.

$$\begin{aligned}
P(Y < 10|\sigma = \sigma_1, \mu = 10.15) &= 0.01 \\
P\left(Z < \frac{10-\mu}{\sigma_1}\middle| \mu, \sigma_1\right) &= 0.01 \\
\left(\frac{10-\mu}{\sigma_1}\right) &= -2.33 \\
\sigma_1 &= .0789.
\end{aligned}$$

We CUSUM the sample variances, and call them X_n. Using Equation 4.3 with $\sigma_0 = 0.05$ and $\sigma_1 = 0.0789$, we see that

$$k = -\frac{2\ln(\sigma_0/\sigma_1)\sigma_0^2\sigma_1^2}{\sigma_0^2 - \sigma_1^2} = .00381.$$

We want our in-control ARL to be 500. Using our program in Figure 4.1, we find that h^+ should be set to 0.0087.

We also find that the ARL until a signal, when σ does shift to σ_1, is 4.8. In other words, the process average 500 rational groups until a false signal when the process is in control. It only average 4.8 rational groups until it signals when it shifts to the out-of-control state at σ_1.

```
Program to find ARL for upward shifts in the standard deviation of a
normal process using a CUSUM of the sample variances.
Copyright 1997, D. M. Hawkins and D. H. Olwell
Enter sigma in control, sigma out of control, sample size, and
desired ARL:
.05 .0789 5 500
data input complete
ref = 3.811E-03
entering main loop
h+= 8.697E-03
ARL: 499.998
ifault = 0
firarl =: 485.034
out of control ARL : 4.788
out of control FIRARL is: 3.432
Another run? (1=yes, 0=no)
```

FIGURE 4.1. Output showing h^+ for the sugar bag example. Notice the output also returns the out–of–control ARL and FIRARL results.

The short ARL until a signal when out of control can be interpreted as meaning that this variance shift is easily detectable using CUSUM methods, although it is not so large that it would be clearly visible without the accumulation done in the CUSUM.

The output in Figure 4.1 includes the FIR results. We see that using a head start of $H = h/2$ results in an decrease of ARL of only 3% when in control, and a 28% faster signal when the process is out of control. This illustrates that in detecting variance increases, just as in detecting changes in normal means, giving the CUSUM a head start can be valuable.

The CUSUM for a downward shift is found similarly.

The combination of the CUSUM for a shift in mean and the CUSUM for a shift in scale provides a complete set of tools for detecting persistent shifts to normal data.

4.1.5 *Shift in the gamma shape parameter* α

The gamma distribution has density function

$$f(y|\alpha, \beta) = \exp\left(-y/\beta + (\alpha - 1)\ln y - \alpha \ln \beta - \ln(\Gamma\alpha)\right) \tag{4.6}$$

The gamma distribution underlies the CUSUM for change in variance: α is one half the number of degrees of freedom of the variance, and $\beta = 2\sigma^2(n-1)$.

We now examine the gamma distribution as a model in its own right, and consider CUSUM schemes for a shift in the first parameter α. Recall that α is a shape parameter. A CUSUM for α is a CUSUM for a change in the shape of the distribution. An increase in α means the distribution has become more bell shaped; a decrease in α means the distribution has become more right–skew.

We also recall that the expected value for a gamma variate, with our parameterization, is given by $E(Y) = \alpha\beta$. So an increase in α also result in an increase in the expected value of the output of the process.

What kind of processes are well modeled by gamma distributions? Since the gamma distribution is a continuous, positive, skew distribution, we consider processes with such outputs. The time needed to complete a task is one example. Such times necessarily must be positive continuous variables, and experience sadly suggests that they are often right-skew, as well. Insurance claims have also been modeled successfully with the gamma distribution (Baxter, Coutts, and Ross, 1980).

The general theory of testing for a change in α is given in Chapter 6, where Equations 6.7, through 6.9 specify the CUSUM design. Like the CUSUM to detect a variance shift, we do not CUSUM Y, but rather a transformed variable:

$$X_n = \ln(y_n).$$

In the context of our task completion times, $T_n \sim \Gamma(\alpha, \beta)$. The statistic we use for the CUSUM is not the time itself, but its natural log, $a(T) = \ln(T)$. The reference value is given by

$$k = \frac{\ln \beta(\alpha_0 - \alpha_1) - \ln\left(\frac{\Gamma(\alpha_0)}{\Gamma(\alpha_1)}\right)}{\alpha_1 - \alpha_0}.$$

We now work an example to show the procedure in action. Assume that the time T to complete the task follows a gamma distribution

$$T \sim \Gamma(\alpha, \beta) \sim \Gamma(4, 1).$$

Suppose we want to monitor for a shape shift, and decide to aim for maximum sensitivity to an increase in α to $\alpha_1 = 5$. We want our in–control ARL to be 500. What should the design of our CUSUM scheme be?

The reference value k and the resulting CUSUM scheme are:

$$\begin{aligned} k^+ &= -\frac{\ln(1)(4-5) - \ln(\Gamma(4)/\Gamma(5))}{4-5} \\ &= \ln(4) \\ S_0^+ &= H^+ \\ S_n &= \max(0, S_{n-1} + \ln(T_n) - k^+). \end{aligned}$$

The only remaining parameter is h, the decision interval value. To find this value, we turn to another of our software programs. The output is shown in Figure 4.2. We see that the value for $h^+ = 3.67$. This completely specifies our CUSUM scheme.

```
Program to find ARL for upward shifts in the
scale parameter of a gamma process using a CUSUM
of the natural logarithm of the individual
observations.
Enter alpha in-control, alpha out-of-control,
beta, and desired ARL:
4 5 1 500
data input complete

ref = 1.386
entering main loop
 .
h+ = 3.671
ARL: 500.017
ifault = 0
firarl =: 457.307
out of control ARL :
27.12
out of control FIRARL is: 16.744
Another run? (1=yes, 0=no)
```

FIGURE 4.2. Output for the upward shift in α example for the Gamma distribution.

The software also reports the ARL of the CUSUM if the process parameter does shift to the target out-of-control level. We note from Figure 4.2 that the speed of detection for our out-of-control state is not as great as we might like: our ARL out of control is 27.12. It is somewhat more difficult to detect the difference between gamma distributions with shape parameters so similar.

The program to find h^+ for the increase in α is slightly complicated by the fact that we are using the CUSUM of a transformed variable, $\ln(T)$. We have adjusted our program and our algorithms to reflect this (sparing you, dear reader, this chore).

A second example for a shift in α for a gamma distribution might come from use of the gamma distribution as a model for the amount of monetary damage in an accident, and the size of the resulting claim. An insurance company might use a gamma model fit on historical data, and be interested in detecting both scale changes and shape changes.

4.1.6 Example — shift in β

Assume that the damage done to an automobile in a collision is expressed in thousands of dollars. Further assume that historical data have shown that the amount of damage to a pool of vehicles of a certain type insured

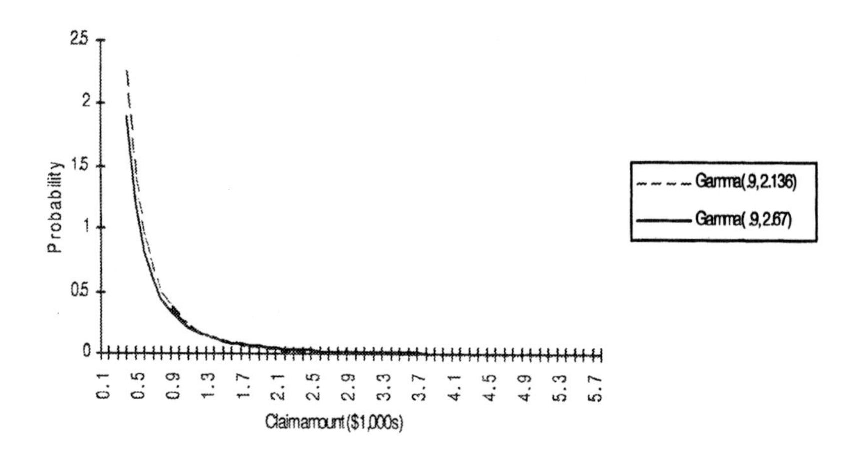

FIGURE 4.3. The two densities for the insurance claim example for a shift in β. The dashed line gives the in-control distribution $\Gamma(0.9, 2.136)$. The solid line gives the out-of-control density for which maximum sensitivity is desired, a $\Gamma(.9, 2.67)$.

by a company has been found to be well modeled by a gamma distribution with parameters $\alpha = 0.9$, and $\beta = 2.136$, or $Y_i \sim \Gamma(0.9, 2.136)$. This implies that the average amount of damage is \$1920, and the most frequent amount of damage is the mode, \$0. These figures are from the unpublished analysis of 103,000 claims from a major automobile insurer for the year 1996. Baxter, Coutts and Ross's British data were compatible with $\alpha = 1$ (an exponential distribution) and so slightly heavier tailed than these figures. The distribution of claims, however, can be much affected by insurance company policies, and so is unlikely to be exactly the same in current US experience as in the older British data reported by Baxter.

The insurance company needs a CUSUM scheme that give it prompt detection of an 25% increase in the scale of the claims. This is equivalent to a shift in β from 2.136 to 2.67. The shift in the densities is illustrated in Figure 4.3.

We CUSUM the claims, Y_i; no transformation is necessary for a CUSUM to detect a scale shift. The optimal reference value k is given (from Equation 6.19) as

$$k = -\frac{\alpha\,(\ln(\beta_0) - \ln(\beta_1))}{\beta_0^{-1} - \beta_1^{-1}} = 2.14485$$

We obtain the decision interval value, $h^+ = 25.1591768$, using our program shown in Figure 4.4. Since the ARL to signal when the process shifts out of control is 71.9, we see that it is relatively difficult to detect a small

```
Program to find ARL for upward shifts in the
scale(beta) parameter of a Gamma (alpha,beta)
using a CUSUM of the observations.


Copyright 1997, D. M. Hawkins and D. H. Olwell


Enter beta in-control, beta out-of-control, alpha, and desired ARL:
2.136 2.67 .9 500
ref= 2.14485582
data input complete


entering main loop
h+ = 25.1591768
ARL: 499.995044
ifault = 4
firarl =: 440.924705
out of control ARL :
71.8792701
out of control FIRARL is: 49.8350595
Another run? (1=yes, 0=no)
```

FIGURE 4.4. Output for the upward shift in β example.

(here 25%) shift in the scale of the claims. This not surprising, given the strong similarity of the density functions in Figure 4.3.

4.2 The inverse Gaussian family

For positive skew data, the inverse Gaussian distribution provides an alternative distributional model to the gamma distribution. Although less well known, the inverse Gaussian distribution is more tractable than the gamma distribution. Of course, when the gamma distribution is known to be the correct model, as in the case of normal variances, it should be used. If there is no *a priori* reason to use the gamma distribution for positive skew data, the inverse gamma distribution should be considered.

We present the details of the use of CUSUM schemes based on the inverse Gaussian distribution in the remainder of this chapter.

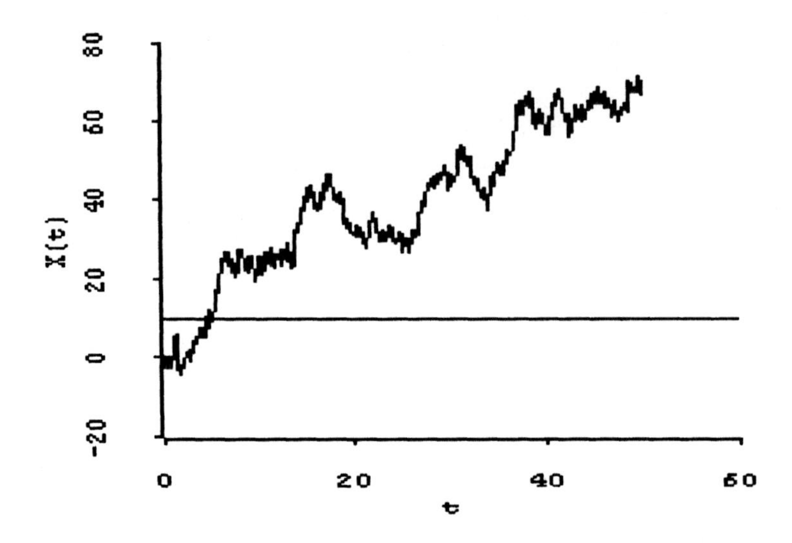

FIGURE 4.5. First passage time illustration for Brownian motion with drift. In this example, we are interested in the distribution of the first time $X(t) > 10$. For this sample path, the time is approximately $t = 6$.

4.2.1 Background

Many industrial problems yield measures with skewed positive distributions: examples are component reliabilities, time to completion of tasks, and insurance claims.

There are already a number of statistical distributions that are positive and right-skew: the gamma, lognormal and Weibull are examples. This raises the question of why another should be needed.

There are three major advantages to using the inverse Gaussian distribution to model skewed data: an appeal to the underlying physical properties of the process being modeled; the notion of failure rates and their asymptotic behavior; and the tractability of the sampling distribution of the inverse Gaussian. We discuss each of the three in turn.

The inverse Gaussian distribution arose in the context of first passage times for a Wiener process. In other words, the inverse Gaussian distribution models the time that it takes a process subject to one-dimensional Brownian motion with drift to reach a certain point. This is illustrated in Figure 4.5. Each possible sample path for the Brownian motion has a first passage time associated with it, and the inverse Gaussian distribution is the distribution of these first passage times. In financial settings, the logarithm of the price of a security is often modeled as a Wiener process with drift. The distribution of the time until a security reaches a certain price is of key interest to those working with derivative securities such as options, and provides a natural setting for the inverse Gaussian distribution.

In an industrial setting, if the underlying process can be thought of as a Wiener process, then the use of the inverse Gaussian seems especially appropriate.

For example, the time to failure of a complex system might be the time taken for many small random degradations to accumulate to the point of causing system failure. This might provide justification *a priori* for trying to model the time to failure as inverse Gaussian. This physical reasoning suggesting the inverse Gaussian distribution is lacking for, say, the log-normal distribution, whose rationale for use would be based on multiplicative effects.

Secondly, consider the failure rate $r(t)$ of a system as a function of the time t. We define

$$r(t) = \frac{f(t)}{1 - F(t)},$$

where $r(t)$ denotes the instantaneous rate of failure for a process conditional on its having lasted a certain time, and $f(t)$ and $F(t)$ are the density and distribution functions for the probability of failure at time t. For a Poisson process with time to failure modeled by the exponential distribution with parameter λ, $r(t) = \lambda$, a constant failure rate.

This assumption of constant failure rate is rather strong. Some applications call for a monotonic failure rate, some for an increasing failure rate (IFR); and others for a decreasing failure rate (DFR). For these it is possible to use the Weibull distribution , with density $f(t, \alpha, \beta) = \alpha\beta^{-\alpha}t^{\alpha-1}\exp(-(t/\beta)^{\alpha})$, $(t, \alpha, \beta > 0)$. Then $r(t) = \alpha\beta^{-\alpha}t^{\alpha-1}$, and is decreasing for $\alpha < 1$ and increasing for $\alpha > 1$.

In many situations that are characterized by a "burn in" process, it seems appropriate to have an initially increasing then decreasing failure rate (Chhikara and Folks, 1977). A positive, initially IFR then DFR process is sometimes modeled by the log-normal distribution. For such a process,

$$r(t) = \frac{f_{\mu,\sigma}(\ln(t))}{t\,(1 - \Phi((\ln(t) - \mu)/\sigma))}, t > 0,$$

where $\Phi(t)$ is the standard normal cumulative distribution function (CDF), and $f_{\mu,\sigma}(t)$ is the corresponding normal density. This failure rate is non-monotonic: initially increasing, then decreasing. However, its asymptote is zero; for many reliability situations, this asymptotic failure rate of zero seems illogical. This suggests that the log-normal model may also be inappropriate.

An alternate model uses the inverse Gaussian process. Its probability density function and failure rate are given by:

$$f(t) = \left(\frac{\lambda}{2\pi t^3}\right)^{1/2} \exp\left(\frac{-\lambda(t-\mu)^2}{2\mu^2 t}\right) \qquad (4.7)$$

$$r(t) = \frac{f(t)}{1-F(t)} \tag{4.8}$$

$$= \frac{\left(\frac{\lambda}{2\pi t^3}\right)^{1/2} \exp\left(\frac{-\lambda(t-\mu)^2}{2\mu^2 t}\right)}{\Phi\left(\sqrt{\frac{\lambda}{t}}\left(1-\frac{t}{\mu}\right)\right) - e^{\frac{2\lambda}{\mu}}\,\Phi\left(-\sqrt{\frac{\lambda}{t}}\left(1+\frac{t}{\mu}\right)\right)}. \tag{4.9}$$

This failure rate is also nonmonotonic, initially increasing to a maximum then decreasing. Its asymptotic failure rate is given by

$$r(t) \to \frac{\lambda}{2\mu^2} \neq 0.$$

This provides a second strong argument for using the inverse Gaussian rather than the log-normal distribution to model lifetimes; – it is hard to conceive of physical processes where the failure rate would decrease to zero as the log-normal model presupposes.

The third argument for using the inverse Gaussian is that the sampling distributions of the maximum likelihood estimators (MLEs) of the parameters are known and easy to use. Using the inverse Gaussian avoids the need to transform the data prior to finding MLEs, as is the case with the log-normal distribution.

The inverse Gaussian distribution is highly flexible for modeling positive skew processes. We illustrate with Figures 4.6 and 4.7. Figure 4.6 shows the density of an inverse Gaussian variate with constant $\lambda = 1$ and various values of μ. Figure 4.7 holds μ fixed and varies λ.

We note that conventional quality control charts motivated by the normal distribution would be very inappropriate here. Because of the heavy tail to the right, charts for centrality never behave as expected, signaling too frequently. This situation argues for inverse Gaussian charts, and illustrates that they are not of only theoretical interest, but may help control these processes better.

The inverse Gaussian distribution $IG(\mu, \lambda)$ with density function given by Equation 4.7 has two parameters, μ and λ. We term μ, the expected value of the distribution, the "location" parameter, and λ the "scale" parameter. (The shape is given by $\phi = \lambda/\mu$, and as $\phi \to \infty$, the distribution becomes symmetric.) The inverse Gaussian is within the univariate exponential family with respect to each of these parameters when the other is assumed fixed, as is standard practice in SPC.

The schemes that follow are for individual observations, or samples of size 1. This has the advantage of maximum flexibility: one can always chart a larger sample as a group of individual observations. These schemes are easily adapted to rational groups of size greater than 1 by making CUSUMs of the $\bar{X}$ and $V = \sum\left(1/X - 1/\overline{X}\right)$, the minimal sufficient statistics for a sample consisting of a rational group of size greater than 1.

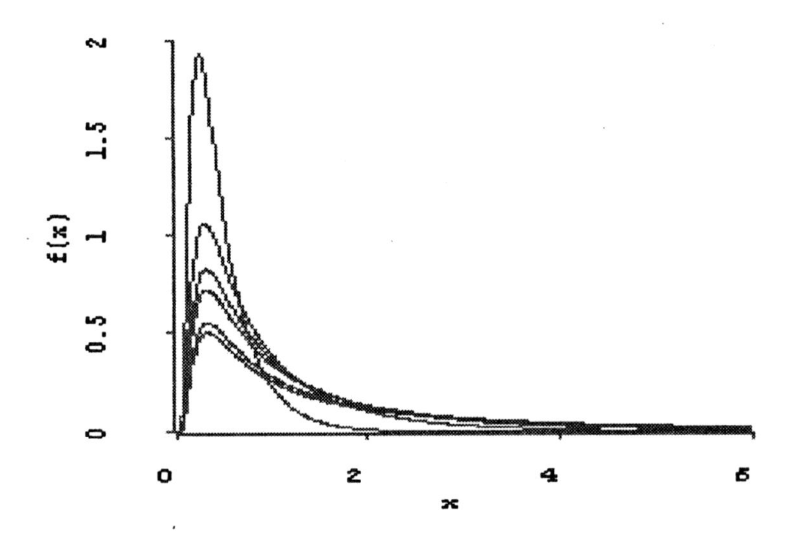

FIGURE 4.6. A sheaf of $IG(\mu, 1)$ densities for $\mu = .5, 1, 1.5, 2, 5,$ and 10

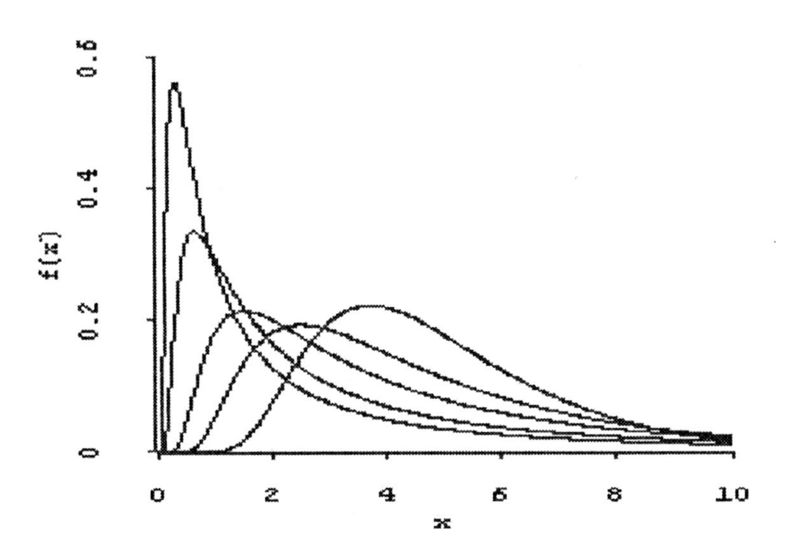

FIGURE 4.7. A sheaf of $IG(5, \lambda)$ densities for $\lambda = 1, 2, 5, 10,$ and 25

The maximum likelihood estimators for μ and λ are known to be

$$\begin{aligned}\hat{\mu} &= \overline{X} \\ \hat{\lambda} &= \frac{n}{V}.\end{aligned}$$

where V was defined in the previous paragraph and n is the number of observations.

4.2.2 *Shift in mean*

Borrowing heavily from material in Chapter 6, with the scale parameter λ fixed, we can write the density of Y in terms of μ as

$$f(y,\mu) = \exp(a(y)b(\mu) + c(x) + d(\mu))$$

with $a(y) = y$, $b(\mu) = -\lambda/2\mu^2$ and $d(\mu) = \lambda/\mu$. Suppose we want to tune the upward CUSUM for a shift to mean μ^+, and the downward CUSUM for a shift to μ^-. Then the reference values k^+ and k^- are given by Equation 6.32, which simplifies to

$$k^+ = \frac{2\mu\mu^+}{\mu+\mu^+} \tag{4.10}$$

$$k^- = \frac{2\mu\mu^-}{\mu+\mu^-}. \tag{4.11}$$

These values for k are the *harmonic mean* of the in-control and out-of-control parameters.

The two-sided CUSUM for a shift in mean is then

$$\begin{aligned}S_0^+ &= H^+ \\ S_0^- &= -H^- \\ S_n^+ &= \max(0, S_{n-1}^+ + Y_n - k^+) \\ S_n^- &= \min(0, S_{n-1}^- + Y_n - k^-).\end{aligned}$$

We provide the usual software routines for finding h^+ and h^- to meet a given ARL.

4.2.3 *Shift in scale parameter*

To derive our CUSUM for the scale parameter, we fix μ and again consider the resulting 1-parameter exponential family, which has

$$a(y) = \frac{(x-\mu)^2}{x\mu^2},$$

$b(\lambda) = -\lambda/2$, and $d(\lambda) = \ln(\lambda)/2$.

μ_0	μ_1	λ	h	ARL in control	ARL out-of-control
3	3.5	5	1	4.742	3.639
			5	16.340	9.730
			10	44.877	20.314
			20	178.354	47.989
			40	1233.208	115.569
μ	λ_0	λ_1	h	ARL in control	ARL out-of-control
3	5	4	5	31.691	17.446
			10	105.302	40.639
			20	532.991	101.779
			40	5163.017	239.129

TABLE 4.1. Some in-control and out-of control ARL values for various CUSUM parameters. Out-of-control values are taken for the parameter at the alternate (tuning) value.

Again looking ahead to results from Chapter 6, (Equations 6.36 through 6.38), the optimal CUSUM for an upward shift from λ_0 to λ_1 is

$$\begin{aligned} a(y) &= \frac{(y-\mu)^2}{y\mu^2} \\ k^+ &= -\frac{\ln(\lambda_0/\lambda_1)}{\lambda_0 - \lambda_1} \\ S_0^+ &= H^+ \\ S_n &= \max(0, S_{n-1} + a(Y_n) - k^+). \end{aligned}$$

On the face of it, this gives yet another new situation for which we need special results, but this turns out not to be the case. It is well known (Chhikara and Folks, 1989) that $\lambda a(y) \sim \chi_1^2$. Accordingly, the CUSUM for a change in λ simplifies to a Gamma CUSUM for a change of scale, which we previously derived.

The reference value k of the CUSUM is, as usual, chosen by consideration of the out-of-control level of λ for which we want maximum sensitivity.

The CUSUM design is then completed by selection of the h value that gives some desired in-control average run length (ARL). Again, we use our programs to simplify the task of finding h. Some values of h and the corresponding ARLs are in Table 4.1.

4.3 Example from General Motors

Desmond and Chapman (1993) examined the task completion times of crews of workers at the General Motors plant in Oshawa, Ontario. They looked at three processes, of which we consider the third, a "radio kit-

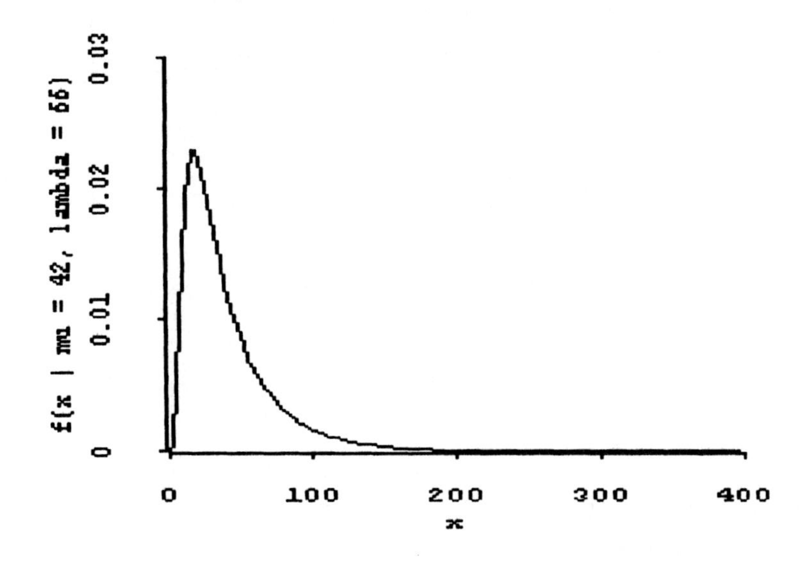

FIGURE 4.8. Density for an $IG(42.6257, 66.282)$ distribution. Note the heavy tail to the right.

ting station", which they describe as well modeled by an inverse Gaussian distribution.

General Motors pre-processed the data prior to releasing it to Desmond and Chapman. The completion times were sorted; accordingly, no analysis depending on the actual time ordering of the original data can be applied. In addition, Desmond and Chapman screened for what they considered outliers (they termed them "bogus readings") and removed them. The article then reported the MLEs from 1,955 observations of this process.

The task completion times are described accurately as inverse Gaussian random variables, with $\hat{\mu} = 42.6257$ and $\hat{\lambda} = 66.282$. The units of time were not specified in the paper, but we assume from the context and the authors confirm in private correspondence that the units are in seconds. Figure 4.8 shows the density of this distribution. We note the very heavy tail to the right. The median for this model is 32.5051, and the mode is 18.1071.

As this is a station in an assembly line, fast steady flow is desirable. Increases in the mean processing time or a decrease in λ, or both, decrease the service rate at this station and can cause slowdown in the overall assembly process. On the other hand, decreases in μ and increases in λ allow management to identify circumstances that reflect improved service rates and decreased variation in the process. Thus increases and decreases in both parameters are of great interest – on one side for detecting problems, and on the other for detecting improvements in the process.

```
This program finds the h value for a CUSUM scheme for an IG(mu,
lambda) to detect an upward shift in mu.

Copyright 1997, D.M. Hawkins and D.H. Olwell
enter mu, mu_o, lambda, and desired arl
42.6257 70 66.282 100
ref = 52.986
data input complete

entering main loop

h+ = 167.449
ARL: 100.002
ifault = 0
firarl =: 91.731
out of control ARL : 10.869
out of control FIRARL is: 8.496
Another run? (1=yes, 0=no)
```

FIGURE 4.9. Output for finding the CUSUM scheme for an upward shift in μ of an $IG(42.6257, 66.282)$ process, tuned to $\mu_1 = 70.0$.

4.3.1 CUSUM chart for location

We now turn our attention to CUSUM charts for μ. We construct a table of ARLs for various shifts in μ, and present it in Table 4.2. The ARLs presented are for a shift to the indicated out-of-control state, with the CUSUM parameters selected for maximum power to detect changes to that state.

Table entries are found using a FORTRAN program. For example, the entry corresponding to an out-of-control mean of 70 is found to have the corresponding CUSUM scheme with parameters $k = 52.9861$ and $h^+ = 167.4488$. A copy of the output is shown in Figure 4.9.

The CUSUM scheme detects these changes much more quickly than the Shewhart scheme presented in Olwell (1996).

4.3.2 CUSUM chart for λ

We now turn our attention to CUSUM charts for λ. We construct a table of ARLs for various shifts in λ, and present it in Table 4.3. The ARLs presented are for a shift to the indicated out-of-control state, with the CUSUM parameters selected for maximum power to detect changes to that state. The ARLs are found using the routines for upward and downward shifts in the scale parameter of a gamma function.

For example, consider the entry for the out-of-control $\lambda_1 = 60$, given in Table 4.3. In control, the distribution was a $\Gamma(.5, 2)$. Out of control, it shifts

ARLs out of control	
μ	*ARL*
20	6.93
25	11.32
30	18.83
35	33.16
40	65.23
42.6257	100
45	64.42
50	34.28
60	16.54
70	10.86
80	8.23

TABLE 4.2. Table of ARLs for CUSUM for the mean, GM example. We use an in-control ARL of 100 for $\mu = 42.6257$ and $\lambda = 66.282$. The table indicates how difficult it is to detect changes of different magnitude in the mean of this very dispersed, very skewed process.

to a $\Gamma(.5, 2\lambda/\lambda_o) = \Gamma(.5, 2.21)$. The parameters for the CUSUM scheme can be found using the FORTRAN routines from earlier in the chapter, and are $k = 1.05, h^+ = 10.79$. In control, the ARL is 100.00, and out of control it is 62.59. The remainder of the table is completed similarly.

Since the variance of $X \sim IG(\mu, \lambda)$ is given by $VAR(X) = \mu^3/\lambda$, we see that as λ decreases, the variance increases. As λ decreases, the distribution also becomes more right-skewed. As we detect shifts in λ, we discover changes in both the shape of the distribution and in the variability of the process. The tables allow us to design our CUSUM for optimal detection of a shift of a certain size, size depending on process knowledge and loss considerations

Detection of increases in λ allow us to identify situations resulting in reduced process variability. This corresponds to process improvement detection, which is an equally valuable result of SPC.

The CUSUM of individual observations does not display increased ARLs for out-of-control states with increased λ, which was a shortcoming of the Shewhart scheme for λ presented in Olwell (1996). These two schemes provide quick response to shifts in μ and λ, allowing quick identification and remedy of *out-of-control* conditions.

4.3.3 Remarks

The inverse Gaussian distribution is not widely used in statistical quality control practice. We have included this example, first because we think that the IG provides a useful alternative for modeling positive skew data, such as the task completion times of this example. and second because working

ARLs out of control	
λ	*ARL*
10	3.05
20	5.79
30	10.27
40	18.05
50	32.52
60	62.59
66.282	100
70	80.82
80	52.60
90	38.92
100	31.06
110	26.01
120	22.51
130	19.95
140	18.06
150	16.50

TABLE 4.3. Table of ARLs for CUSUM for λ, GM example. We use an in-control ARL of 100 for $\mu = 42.6257$ and $\lambda = 66.282$. Note that this is a CUSUM of individual observations.

with a novel distribution allows us to demonstrate the general applicability of the CUSUM methodology developed in Chapter 6 to all members of the exponential family.

4.4 Comments

In this chapter, we have considered CUSUM schemes for the gamma and inverse Gaussian distributions. These schemes allow CUSUM charts to be constructed for positive skew data.

> The software and schemes provided in this text allow the quality practitioner to have real choices among possible models when designing statistical process control schemes.

4.5 Further reading

Gan and Choi (1994) give a fast FORTRAN code (based on the explicit solution provided by Vardeman and Ray 1985) to find the ARLs of data following the exponential distribution.

We mentioned the Weibull distribution in passing. This brief mention underplays the fact that the Weibull distribution is the distribution used most widely for modeling life data. The ability to control Weibull data is therefore particularly important for manufactured devices that wear out, a range covering such diverse products as pacemakers, diesel truck engines, and video displays.

Our apparent omission of the Weibull is not serious, however, because the Weibull distribution can be handled with an easy adaptation of techniques for the exponential distribution; Johnson (1966) provides a fuller discussion. The Weibull distribution has density

$$f(y, \alpha, \beta) = \alpha \beta^{-\alpha} y^{\alpha-1} \exp(-(y/\beta)^{\alpha})$$

and so if we transform our lifetime reading Y_n to $X_n = Y_n^{\alpha}$, we can easily see that X_n follows a regular exponential distribution with mean β^{α}. CUSUM monitoring of X_n can then be used to check for changes in the mean. This approach assumes that when the process goes out of control, β changes but α does not, and on this assumption we can use properties of exponential CUSUMs to calculate the out-of-control ARL of the CUSUM. It is possible for the process to go out of control because of changes in α (which determines the shape of the distribution). The CUSUM of X_n is not optimal for shifts in α, but since changes in β are usually much more important than changes in α, and since the CUSUM of X_n does have some ability although less than optimal, to detect changes in α, you usually get by using just the X_n CUSUM.

Another paper by Johnson (1963) discussed CUSUM control of the folded normal distribution (the absolute value of a zero-mean normal variable). We do not know of any common real-world situations where this distribution is required, although it might be used for a CUSUM of moving ranges.

In a similar vein, Hawkins (1992) included a CUSUM of sample ranges of rational groups of normal data as an example of a possible CUSUM.

5
Discrete data

> The ants go marching one by one, hurrah, hurrah ...
> *American camp song*

Introduction

The data we have been considering so far were on the continuous scale. In classical SPC terminology they are "variables" measures. The other classical SPC data type is "attribute" data , originally counts of the number of good and of defective (or in current terminology "conforming" and "nonconforming" items) in a sample. We draw the distinction a little differently: into the standard statistical distinction between continuous and discrete measurements.

5.1 Types of discrete data

Discrete data form an important area of SPC applications, arising in several contexts. Within this, there are two broad categories of data: those that arise from categorization, and those that arise from counting. We call these two classes "binomial" data and "count" data, labels that should not be taken to imply more than this broad grouping.

5.1.1 Binomial data

In some settings items can be classified as either having or not having some attribute. For example, surgery patients may be categorized by whether they did or did not have postoperative wound infections, or a TV may be classified as having either acceptable color or unacceptable color. We group this class of data together under the heading "binomial" data, or "yes/no" data.

Some measures are inherently binomial. For example, a taster may judge a sample of cheese to be either acceptable or unacceptable. Sometimes a measure is not inherently binomial, but comes about from categorizing some other numeric measure. For example, library books may be classified by whether they were returned within seven days of borrowing; this is a binomial measure found by categorization based on grouping the numeric measure of how many days a book was out.

As another example, consider again our sugar bag process. The fill weight of bags — the weight of sugar put into each bag — is very important. If the bags are underweight, the supplier risks severe legal penalties. If the bags are overweight, the extra product given away cuts into profits. There are two ways of monitoring the fill weight. One is to take a sample of the filled bags and weigh them accurately. This provides a continuous "variables" measure that can be assessed using Shewhart and CUSUM charts of the type we discussed in earlier chapters. Another approach is to use a scale that just check whether the sampled bag's weight is above or below some set point. Each sampled bag will then be categorized as "go" or "no-go."

This type of control was first developed for dimensional checking. To check the diameter of manufactured bolts, a manufacturer could have inspectors carefully measure the diameters of sampled bolts with calipers. A much simpler and faster approach is to make two templates so that any bolt that could not be put through the hole in the larger template was too large, and any bolt that could be put through the hole in the smaller template was too small. Bolts that go through the larger but not the smaller template are classified as acceptable. Generations of American soldiers have worked with gauges of this type as they set headspace and timing on the .50 caliber machine gun.

The attraction of these "gauge" measures is that they are quick and easy. Their drawback — that they provide much less information than variables measurements, and therefore less effective controls — can be removed by careful choice of the gauges, and by using larger samples than would be used for variables measures. If a gauge classification costs one tenth as much as a variables measure and three times as many items need to be sampled to get the same level of process control, then the gauge approach is a clear winner.

5.1.2 Count data

The second type of discrete data is "count" data. For example, in an auto assembly line each body sampled for control on the paint operation may be studied for surface blemishes and a count made of the number of blemishes. In manufacturing magnetic tapes, the tapes are studied for patches of imperfect magnetic medium. A sampled reel of tape then provides a count of the number of imperfections. This is an update of the historical example in fabric weaving, where a sampled bolt of fabric have a count of the number of weaving flaws.

The difference between the second type of data, count data, and the binomial type is that the count on any sampled item can in principle be any nonnegative integer. In classification, the sampled item either does or does not have the characteristic. In a collection of m sampled items, binomial data would necessarily be in the range $0, \ldots, m$. By contrast, the number of bad spots in a magnetic tape has (at least in principle) no upper bound.

Shewhart control charts of discrete data are standard tools. Binomial data are handled using the p and the np charts, count data with the c or u charts. We develop parallel CUSUMs for detecting persistent rather than isolated shifts in the parameters of discrete data.

5.2 The graininess of the ARL function

CUSUMs of an integer-valued X are defined in exactly same way as with variables data. So a V-mask form CUSUM is defined by

$$\begin{aligned} C_0 &= 0 \\ C_n &= C_{n-1} + (X_n - \mu), \end{aligned}$$

where μ is the in–control true mean of the X_n. A decision interval CUSUM for an upward shift is defined by

$$\begin{aligned} C_0^+ &= H^+ \\ C_n^+ &= \max(0, C_{n-1}^+ + X_n - k^+), \end{aligned}$$

where k^+ is the reference value of the upward CUSUM and the head start value H^+ may be zero or nonzero. A value $C_n^+ \geq h^+$, where h^+ is the upward decision interval, signals an upward shift in mean. Downward CUSUMs are defined in the same way.

Since X_n can only take on integer values, the possible values of the CUSUMs C and C^+ can also have a restricted range. The V-mask form CUSUM at stage n has a value that is necessarily some integer minus $n\mu$.

In particular, if μ is an integer (although there is no need for it to be integer-valued), then C_n is also an integer for all n.

Similarly, the decision interval CUSUM's values are necessarily of the form "head start (if nonzero) plus integer minus integer multiple of k^+." This has important implications for CUSUM design. If H^+ and k^+ are integers, then so is C_n^+ for all n, which necessarily takes on one of the limited set of values $0, 1, 2, ..., h^+ - 1$. The decision interval h^+ may as well be an integer, since it is only integral values that can be attained.

> This limited set of possible values of h^+ means that there is a similarly limited set of possible ARLs.

The programs ANYARL and ANYGETH from the Web site deal with the most common needs for CUSUMs of discrete data from the binomial, Poisson, and negative binomial distributions. Each routine calls for the parameters of the data distribution and the CUSUM reference value. ANYARL returns the ARL for a given h, and ANYGETH attempts to find a suitable h for the target ARL.

Consider a CUSUM of Poisson-distributed data (we talk about the Poisson in more detail in the following) with true mean $\mu = 4.3$, and suppose we want to design the CUSUM for an in-control ARL of 1,500 with a reference value of $k^+ = 7$. Supplying this information to ANYGETH produces the following output. In the earlier chapters, we deleted some of the output ANYGETH produces in its search for the solution; here we show it all as it adds some insights.

```
 Enter the Poisson mean (must be positive)
 4.3

What is the Winsorizing constant?
1000

Do you want zero-start (say Z) or FIR (say F)?
 z

Enter k and ARL (zeroes stops run)
 7 1500
k    7.0000 h  3.00000 arls      67.1      65.7
k    7.0000 h  6.00000 arls     911.7     895.8
k    7.0000 h  9.00000 arls   13940.6   13892.8
k    7.0000 h  8.00000 arls    5606.0    5560.3
k    7.0000 h  7.00000 arls    2255.3    2238.1

k  7.0000 h  6.0000 ARL    911.67
          h  7.0000 ARL   2255.33
```

This list covers integer values of h from 6 through 9, the only attainable values for C_n^+ since $H^+ = 0$ and k^+ is an integer. None of these attainable in-control ARLs is close to the target; there is no way we can get an ARL of 1,500 for this μ, k pair. ANYGETH ends by reporting the two h^+ values that straddle the target.

If we really want an ARL close to 1500, this situation is quite disappointing. However, unlike the mean (over which we have no control), we can pick the reference value k^+. If k^+ is a half-integer, then C^+ can take on half-integer as well as integer values, and the list of possible h^+ also expands to include half-integer values. Requesting the CUSUM with $k^+ = 7.5$ and ARL of 1,500, ANYGETH returns with:

```
Enter k and ARL (zeroes stops run)
7.5 1500

k   7.5000 h  3.00000 arls    161.4    157.0
k   7.5000 h  6.00000 arls   2688.8   2669.8
k   7.5000 h  4.50000 arls    496.3    490.1
k   7.5000 h  5.50000 arls   1364.0   1355.2

k  7.5000 h  5.5000 ARL   1364.03
          h  6.0000 ARL   2688.75
```

The choice $k^+ = 7.5, h^+ = 5.5$ comes reasonably close to the target of 1,500. If we wanted more flexibility in the attainable ARLs, we might go a stage further and pick k^+ to be an odd multiple of 0.25, so that quarter-integer values of h^+ also become possible. Some ARLs with $k^+ = 7.25$ given by ANYARL are:

```
k=  7.2500  h=  4.7500 regular ARL    489.47 FIR ARL    482.16
k=  7.2500  h=  5.0000 regular ARL    920.14 FIR ARL    905.07
k=  7.2500  h=  5.2500 regular ARL    941.04 FIR ARL    927.87
k=  7.2500  h=  5.5000 regular ARL   1017.18 FIR ARL   1002.81
k=  7.2500  h=  5.7500 regular ARL   1327.43 FIR ARL   1318.07
k=  7.2500  h=  6.0000 regular ARL   2294.97 FIR ARL   2277.60
k=  7.2500  h=  6.2500 regular ARL   2379.60 FIR ARL   2361.88
k=  7.2500  h=  6.5000 regular ARL   2659.66 FIR ARL   2639.68
k=  7.2500  h=  6.7500 regular ARL   3672.30 FIR ARL   3645.74
```

This shows how making k^+ an odd multiple of 0.25 brings a yet wider range of ARLs into the picture.

This list of the ARLs for different h^+ shows something else strange. In the case of continuous data, the ARL varies quite smoothly with h. This is not at all the case here. As h goes from 4.75 to 5, for example, the ARL increases by 431, but as h goes up to 5.25, the ARL increases by only 21. The next successive increases in ARL in response to the 0.25 increments in h are 76, 310, 967, and 85, respectively. These differences in ARL are very unstable. The reason is easy to see. The probability that a Poisson with mean $\mu = 4.3$ be 13 or more is 0.0005, or 1 in 2,000. If any single X_n is 13 or more, then regardless of where the CUSUM was prior to this large value, it cross any decision interval of $13 - 7.25 = 5.75$ or less. A substantial portion of the false alarms from the values $h = 5, 5.25, 5.5$, and 5.75 comes from this 1 in 2,000 chance of tripping the CUSUM with a single X value of 13 or more. Once we go to $h^+ \geq 6$ a single value of 13 is no longer enough to trigger the CUSUM from a zero start and the CUSUM jumps substantially.

The graininess of discrete data affects ARLs, causing unique problems for the design of discrete CUSUMs.

5.3 The Poisson distribution and count data

In the introduction, we distinguished binomial-type data from count data. The most common statistical model for count data is the Poisson distribution. The Poisson distribution with mean μ has a probability function

$$p(x) = \frac{e^{-\mu}\mu^x}{x!}, x = 0, 1, 2, \ldots$$

The Poisson distribution arises theoretically from the statistical model of a *Poisson process*. An example of this would be fabric weaving.

Suppose that in any short stretch of woven fabric the probability of a flaw is λt where t is the (short) length of fabric being checked; that the occurrence of the flaw is independent of the presence or absence of flaws in other nearby portions of the fabric; and that having two or more flaws in this short length of fabric has a probability of the order t^2. Then the probability of x flaws in a total length T is Poisson with mean $\mu = \lambda T$. This is defined by the probability function

$$p(x) = \frac{e^{-\lambda T}(\lambda T)^x}{x!}$$

The Poisson process has been used as a model for many similar situations of events in a continuum; for example:

- the occurrence of industrial accidents (here T would represent time),
- defects in semiconductor wafers (here T would represent the area of the wafer),
- typesetting errors in printers' galleys (here T would represent the length of the galley checked).

It can also be tried as a potential model for count data generally, even when there is no underlying continuum in which the Poisson events occur. Thus the Poisson can be used to model the number of visible flaws in the paint finish of an automobile, or the number of goals scored in a soccer match, even though the strong assumptions of the Poisson process are not obviously true.

In these situations, the fact that the Poisson process *might be* a reasonable model does not automatically imply that it *is* a reasonable model. Thus industrial accidents turn out not to be too well modeled by the Poisson distribution, but rather to show clustering. Defects in magnetic tapes also tend to occur in patches and so to show wider spread than the Poisson model allows. These realities do not invalidate making good use of the Poisson model but they do warn against its automatic use without checking.

We show some ARL calculations later to illustrate the impact of using the Poisson model when it is inappropriate.

5.3.1 Useful properties of the Poisson distribution

Two important basic properties of the Poisson distribution first relate its mean and variance; and second relate to the sum of independent Poisson variables.

> First: the variance of a Poisson variable equals its mean.

Working in terms of the original Poisson process, the true mean is λT; the variance is also equal to λT.

This fact gives a valuable check on whether a particular set of data is compatible with a Poisson model; if the sample variance is close to the mean, this indicates that the Poisson model is plausible; but if the variance is much greater than the mean, the Poisson model is implausible. Count data whose variance is larger than the mean are said to be *overdispersed.* This overdispersion leads to CUSUMs having shorter (and possibly much shorter) ARLs than Poisson theory would suggest. This means that if you apply a Poisson CUSUM to count data without checking that the distribution is reasonably Poisson you could get a false alarm rate much higher than you were anticipating. This is not only true of the CUSUM though; if you use a conventional Shewhart c chart on data that are overdispersed, that also give more false alarms than you would expect.

> Second: the sum of independent Poisson variables is Poisson.

If X is Poisson with true mean μ_x and Y is Poisson with true mean μ_y and X and Y are independent, then their sum $X+Y$ is Poisson with true mean $\mu_x + \mu_y$.

This result has a number of implications in quality work. One is that if different quality deficiencies follow Poisson distributions and are independent, then they can be added together to form another Poisson quantity.

Let's assume the number of open solder joints in a circuit board is Poisson and the number of cracks in the base is an independent Poisson. Then the total of the open solder joints plus the cracks in the base is another Poisson variable with mean parameter equal to the sum of the mean parameters for open solder joints and for cracks in the base.

This particular illustration is implausible to the extent that it might not make any real sense to add together the number of open solder joints and the number of cracks in the base. Since these are quite different quality problems with different causes, the only time they might reasonably be added is in problems where it makes sense to accumulate many individual potential problems into an overall total problem count.

The other implication is a more useful one. It is that the "base" of the control scheme can be chosen without changing the fundamental properties of the scheme.

For example, open solder joints in printed circuit boards should be rare, so that a count of the number in an individual board, if it followed a Poisson distribution, would have a small mean and so would nearly always be zero. Adding together the total number of open solder joints in 100 sampled circuit boards, however, would give a measure that also followed a Poisson distribution, but with a mean 100 times larger, and so the counts coming from this inspection scheme should be much more discriminating than those given by individual boards.

5.4 The Poisson and CUSUMs

For discrete data, the CUSUM remains the optimal test for detecting a persistent shift in the process parameters, as we show in Chapter 6. So, if you have a count X_n following a Poisson distribution and want to check for a persistent step shift in mean from an in-control level μ_0 to an out-of-control level μ_1, then the optimal test is a CUSUM.

5.4.1 Design for an upward shift

If $\mu_1 > \mu_0$ so that the change being monitored is an increase in mean, then the CUSUM is given by Equations 6.24 and 6.25:

$$\begin{aligned} C_0^+ &= H^+ \\ C_n^+ &= \max(0, C_{n-1}^+ + X_n - k^+), \end{aligned}$$

where the reference value k^+ is defined by

$$k^+ = \frac{\mu_1 - \mu_0}{\ln \mu_1 - \ln \mu_0}.$$

In the normal-mean case, the reference value k was midway between the in-control and out-of-control means for which the CUSUM was being tuned, but as with the gamma and inverse Gaussian, this is also not quite the case in the Poisson CUSUM.

It is true, but not trivially obvious, that k^+ lies between μ_0 and μ_1, lying a bit closer to μ_0 than to μ_1. For example, if $\mu_0 = 4$ and $\mu_1 = 8$, then $k^+ = (8-4)/(2.079 - 1.386) = 5.771$ rather than $(4+8)/2 = 6$.

It is common practice and sensible to round the value of k^+ to some more convenient value. There are two common reasons for wanting to use a rounded value of k^+, and both are matters of convenience. Since the decision interval CUSUM values are either zero or some value of the form *integer minus integer multiple of* k^+, rounding k^+ also rounds the values the CUSUM can take, which can simplify maintaining it (if this is done by hand rather than on a computer) and reading it.

The second reason relates to calculating the ARL using the Markov chain method discussed in Chapter 6. The computational effort and the numerical accuracy of the calculated ARL are both very much dependent on the number of states the Markov chain has. For a CUSUM of a discrete random variable, this number is the number of attainable distinct values in the range $[0,h^+]$. By limiting the number of attainable values of the CUSUM, we can speed up calculation of the ARL.

For example, if $h = 6$ and k is an integer, then the CUSUM C_n^+ has only 5 possible values: 0, 1, 2, 3, 4, 5. If we used the four significant figure value 5.771 for k, the CUSUM would have some 6,000 possible values and calculating the ARL exactly would become a computational nightmare.

For both reasons, the theoretically optimal reference value 5.771 would in practice be rounded to a value like 5.5, 5.75, or 6. Any of these choices could be rationalized as tuning the CUSUM to a slightly different target than that originally intended.

5.4.2 *Downward shift*

The CUSUM for a downward shift ($\mu_1 < \mu_0$) is

$$\begin{aligned} C_0^- &= -H^- \\ C_n^- &= \min(0, C_{n-1}^- + X_n - k^-), \end{aligned}$$

where $k^- = (\mu_0 - \mu_1)/(\ln \mu_0 - \ln \mu_1)$ is the (positive) reference value and the CUSUM signals if $C_n^- < h^-$.

Some workers dislike CUSUMs that take on negative values. If the CUSUM is to be defined to take on only positive values, then reversing all signs, it can be written

$$-C_n^- = \max(0, -C_{n-1}^- + \{-X_n\} + k^-)$$

where the summand $(-X_n)$ takes on only negative values, being the negative of a Poisson random variable and, since C_n^- is zero or negative, the CUSUM of $(-C_n^-)$ is nonnegative.

5.4.3 ARLs

Tabulating the ARL of the Poisson distribution is not as easy as it was for the normal and the gamma distributions. In these continuous members of the exponential family, all random variables can be linearly transformed into a standard form. In the normal case, this standardization allows the entire family of normal distributions to be handled just by tabulating the ARL of the standard normal $N(0,1)$. The ARL of a CUSUM of normal quantities with any mean and any standard deviation can be found from ARLs of the $N(0,1)$.

With the gamma distributions, one cannot get quite this level of compactness, but the scale parameter can be transformed out of a problem, leaving just the shape parameter requiring separate ARL tables for different shape parameters.

Discrete distributions do not have this same possibility of a transformation to a standard form; each value of the parameters must be tabulated separately. For this reason, tables of Poisson ARLs are of very limited potential and calculation using software is a practical necessity for other than simple example calculations.

5.4.4 Example

As an example of CUSUM design and construction, let's consider a fabric weaving process in which we count the number of visible weaving flaws X_n in a standard sampled length of fabric. The in-control mean is $\mu = 5$. We wish to design a combined CUSUM scheme to detect both increases in μ (which would be urgent indicators of quality deterioration) and decreases (which would validate refinements made to the process to try to reduce the number of weaving flaws).

We design the CUSUM for an increase from $\mu = 5$ to $\mu = 8$. This gives the reference value

$$k^+ = \frac{8-5}{\ln 8 - \ln 5} = 6.38.$$

We round this to $k^+ = 6.5$, a change that simplifies the CUSUM itself (by making all values integers or half-integers) and also simplifies the ARL calculations by reducing the number of possible states in the Markov chain. This is equivalent to choosing $\mu_1 = 8.27$ instead of our original choice of 8.

Running the values $k = 6.5, ARL = 1000$ through ANYGETH gives

```
k   6.5000 h  3.00000 arls     20.1     18.0
k   6.5000 h  6.00000 arls     98.6     92.0
k   6.5000 h  9.00000 arls    466.2    446.8
k   6.5000 h 14.00000 arls   5902.4   5822.8
k   6.5000 h 11.50000 arls   1669.6   1635.1
k   6.5000 h 10.50000 arls   1006.4    980.2
k   6.5000 h 10.00000 arls    777.5    751.9

k  6.5000 h 10.0000 ARL     777.45
          h 10.5000 ARL    1006.35
```

An in-control ARL of almost exactly 1000 can be achieved with $k^+ = 6.5$, $h^+ = 10.5$. On the assumption that this ARL is suitable for this problem in this plant, we can then go ahead with performance calculations.

Running the ANYARL code with different μ values and this choice of k, h gives

```
Enter the Poisson mean (must be positive)

6
k=  6.5000   h= 10.5000 regular ARL     63.23 FIR ARL     54.35
7
k=  6.5000   h= 10.5000 regular ARL     15.11 FIR ARL     10.80
8
k=  6.5000   h= 10.5000 regular ARL      7.46 FIR ARL      4.84
9
k=  6.5000   h= 10.5000 regular ARL      4.92 FIR ARL      3.10
10
k=  6.5000   h= 10.5000 regular ARL      3.71 FIR ARL      2.32
```

The response to these increases in mean is quite fast. The targeted shift to $\mu = 8$ can be detected in an ARL of 7.5 readings, and even the much harder target of $\mu = 6$ can be hit in an ARL of 63 samples.

For the downward portion of the scheme, we design for $\mu = 4$. The reference value for this shift is given by

$$k^- = \frac{4 - 5}{\ln 4 - \ln 5} = 4.481$$

which we round to $k^- = 4.5$. This shifted mean of 4 is much closer to the in-control level than was the out-of-control value we used for upward shifts. This reflects the fact that in a well-established process it is not easy to make large improvements but it is very possible to make things much worse.

The ANYGETH code for the ARL of Poisson CUSUMs for downward shifts in mean then gives the following.

```
 Enter the Poisson mean (must be positive)
 5
What is the Winsorizing constant?
1000
Do you want zero-start (say Z) or FIR (say F)?
z
Enter k and ARL (zeroes stops run)
4.5 1000
k    4.5000 h  3.00000 arls        8.7        6.8
k    4.5000 h  6.00000 arls       28.3       22.9
k    4.5000 h  9.00000 arls       71.5       60.0
k    4.5000 h 14.00000 arls      260.5      230.5
k    4.5000 h 19.00000 arls      832.4      767.3
k    4.5000 h 24.00000 arls     2523.6     2394.9
k    4.5000 h 21.50000 arls     1455.2     1369.0
k    4.5000 h 20.50000 arls     1165.2     1090.1
k    4.5000 h 20.00000 arls     1042.1      967.0
k    4.5000 h 19.50000 arls      931.6      866.5

k  4.5000 h 19.5000 ARL     931.60
          h 20.0000 ARL    1042.10
```

so that an in-control ARL very close to 1,000 can be had using $k^- = 4.5$, $h^- = 20$. The performance of this CUSUM can be calculated by supplying a selection of changed means to ANYARL. For example, we may try:

```
 Enter the Poisson mean (must be positive)
 4.5
 k=  4.5000  h= 20.0000 regular ARL    109.56 FIR ARL      81.02
 4.0
 k=  4.5000  h= 20.0000 regular ARL     35.67 FIR ARL      21.24
 3.5
 k=  4.5000  h= 20.0000 regular ARL     20.00 FIR ARL      10.99
 3.0
 k=  4.5000  h= 20.0000 regular ARL     13.80 FIR ARL       7.40
```

The target improvement of $\mu = 4$ can be detected with an ARL of 36. We emphasize that this is a small change in relation to the background variability of the Poisson, so this ARL is perhaps shorter than you should expect.

Figure 5.1 illustrates the operation of this CUSUM. The values are all half-integers, giving the CUSUM the characteristically jagged look associated with discrete data. Midway through the sequence, the mean apparently shifts upward, and the CUSUM crosses its decision interval with a value of 11 at sample number 58. The last observation on the axis was sample number 50. This gives the graphical estimate of the new mean as

$$\hat{\mu}_1 = 11/8 + 6.5 = 7.875.$$

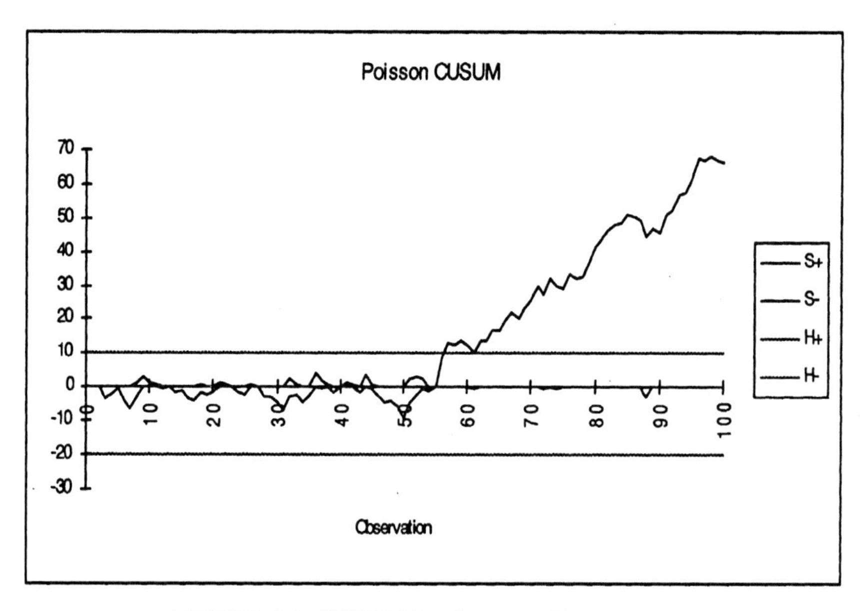

FIGURE 5.1. CUSUM for change in Poisson mean.

The downward CUSUM stayed well above its decision interval for the entire length of the chart, indicating that there was no evidence of a decrease in mean.

5.4.5 The effect of departures from Poisson

As we said at the start of this chapter, the Poisson distribution is a reasonable distributional model to try when faced with count data where the Poisson process looks plausible. *It is not, however, guaranteed to fit count data*, and so it is interesting to know how departures from the Poisson distribution might affect the ARL of the CUSUM designed using the Poisson model.

The negative binomial distribution provides a useful route to investigate this. We say some more later about the use of the negative binomial distribution in its own right. For now, we just use it as a way to move smoothly away from the Poisson distribution into distributions with heavier tails.

There are two parameterizations of the negative binomial distribution. The usual form measures the number of trials until the kth success of a Bernoulli process. The second form measures the number of failures until the kth success. We use this second form, since it allows the response to assume all nonnegative integer values and is more useful for modeling. We use a parameterization that supports asymptotic convergence to the Poisson distribution. This form of the negative binomial distribution has the probability mass function

k	h	σ^2			
		8.00	5.53	5.04	4.3
4.5	32.0	284	555	678	989
5.0	16.0	138	375	518	989
7.5	5.0	66	239	380	1023
10.5	1.5	52	167	251	600
10.5	2.0	86	363	608	1843

TABLE 5.1. Table illustrating effects of departures from Poisson distribution on ARL. μ equals 4.3. Under the Poisson model, $\sigma^2 = 4.3$ as well. That is the right-hand column. The other columns are for negative binomial data with the same mean, but increased variance, reflecting overdispersion. Note the strong reduction in ARL for even modest increases in variance.

$$f(x) = \frac{\Gamma(r+x)c^r}{x!\Gamma(r)(1+c)^{r+x}}$$

($\Gamma(.)$ being the gamma function.) Its mean is $\mu = r/c$ and its variance is

$$\sigma^2 = \mu(1 + 1/c).$$

The Poisson distribution has variance equal to mean, but in the negative binomial distribution the variance is larger than the mean. This overdispersion is determined by the parameter c. If c is near zero then there is a large overdispersion, while large c values lead to near-Poisson behavior of the negative binomial.

The negative binomial turns out to arise, among other ways, when a Poisson process has a λ that is not constant, but is a random variable. In this genesis, the parameter c measures the variability in λ, a large value of c corresponding to a nearly constant λ, and therefore to an X that is very close to Poisson. A small c means a highly variable λ, and a resulting distribution for x that is very different than the Poisson.

Earlier we sketched out some designs for a Poisson with mean $\mu = 4.3$. Let's see what the ARL is if the actual distribution is negative binomial rather than Poisson. Table 5.1 was computed using the CUSARL routine of the Web site, and shows the true ARL for negative binomial data with the same mean $\mu = 4.3$ but different variances and therefore different degrees of overdispersion. The rightmost column is for Poisson data, where $\sigma^2 = \mu = 4.3$.

This table is quite sobering. In all cases, the ARL of the negative binomial data is much smaller than that (rightmost column) of Poisson data with the same mean. Even the column with $\sigma^2 = 5.03$, a variance just 17% larger than the mean (and a level of overdispersion quite hard to detect even in quite large samples) gives ARLs as little as one third of the value they would have with truly Poisson data. The situation with seriously overdispersed data (for example, the first column, where $\sigma^2/\mu = 1.9$) is far worse. Another feature of the table is the impact of the choice of h and k. Small k values (and therefore large h values) can remove a lot of the effect of the overdispersion. This parallels what we saw when CUSUMming nonnormal data using h values based on the normal. The basic reason is similar: with larger h values it is increasingly unlikely that the CUSUM be tripped by a single observation far out into the tail of the distribution.

> The ARLs for CUSUMs of Poisson variables are **very** sensitive to departures from the assumed Poisson distribution.

5.4.6 Checking conformity to the Poisson model

Clearly the tail behavior of the count data is a matter of considerable importance as it can change the ARL dramatically. This suggests that you should check count data for Poisson distribution when using these Poisson charts, if you want to be sure your actual ARL is at least in the ball park of your design goals.

An important characteristic of the Poisson distribution is that its true variance equals its true mean. An easy check that can be applied to count calibration data is to see whether this equality is plausibly present in the calibration data. If so, then you may proceed with caution using the design based on the Poisson distribution; if not (and in particular if the variance seems to be larger than the mean, as is often the case) then you should anticipate that the Poisson ARL calculations may be badly misleading.

The check on the equality of the mean and variance is the "Poisson dispersion test." It consists of gathering the calibration sample of size m, say, and computing their sample mean $\overline{X}$ and sample variance s^2. The formal test for overdispersion then is made by comparing the value of $(m-1)s^2/\overline{X}$ with tabled values of the χ^2 distribution with $m-1$ degrees of freedom.

Suppose that a calibration sample of size $m = 50$ yields a sample mean of $\overline{X} = 4.3$ and a sample variance of $s^2 = 6.6$. Then the dispersion test statistic is

$$\frac{49 \times 6.6}{4.3} = 75.2.$$

This value is above 74.9, the upper 1% point of a χ^2 with 49 degrees of freedom, and so the data seem to suffer from overdispersion. Using Poisson design calculations understate the actual false alarm rate that this process is likely to produce.

Sometimes the model for the negative binomial distribution as a Poisson distribution whose λ is not constant will resolve overdispersion situations like this. In particular, if the calibration data were gathered from a time period in which λ was not constant — that is, in which the process was not in statistical control — then overdispersion is likely to be present. Dissecting the calibration data may then isolate periods in which the true mean was at different values. A V-mask form CUSUM is an excellent diagnostic for this.

We discuss the CUSUMs for the negative binomial distribution at the end of this chapter. If we reject a Poisson model for our data because of over dispersion, we can try a negative binomial model with parameters:

$$c = \frac{\mu}{\sigma^2 - \mu} \tag{5.1}$$

$$r = \frac{\mu^2}{\sigma^2 - \mu} \tag{5.2}$$

where we may use large sample estimates for the actual parameter values, if necessary.

> Departures from the Poisson model affect the CUSUM. One should test for overdispersion prior to adopting a Poisson CUSUM. If overdispersion exists, a negative binomial model may be a better choice.

5.5 Weighted Poisson CUSUMs

In the discussion of the Poisson distribution so far, we assumed that all the samples taken were of the same size. We might take a sample of 50 printed circuits from a production line and count the total number of weak solder joints. If the number of weak solder joints per board follows a Poisson distribution with parameter λ, then the total number in the sample of 50 follows a Poisson distribution with parameter $\mu = 50\lambda$.

In sampling magnetic tape for surface flaws, the sampling scheme might use a standard length of 500 feet. If the flaws occur according to a Poisson process with a mean occurrence rate of λ per foot, then the number of flaws in the sampled length is Poisson with parameter $\mu = 500\lambda$.

In both cases, a standard Poisson distribution arises.

Sometimes, it may make more sense to vary the size of the sample taken. It may be easier to take smaller samples from shorter production runs and larger ones from longer runs. The reasons for this are generally not statistical (the optimal sample size being hardly affected by the population size) but relate more to the greater administrative ease of taking a sample when its size is allowed to vary a bit.

To bring these problems into a common framework, suppose that the size of the nth sample is W_n, and that the process is Poisson with a parameter

λ. In the circuit board example, λ would be the average number of weak solder joints per board, and W_n would be the number of boards taken in the nth sample. In the magnetic tape example, λ would be the mean number of flaws per foot, and W_n would be the number of feet sampled from the nth sample tape.

Then X_n be Poisson with mean $\mu = W_n\lambda$. If the W_n are unequal, then the X_n have different means, making it unclear how to construct the CUSUM. Looking at the decision interval form helps resolve this difficulty. Rather than the mean μ, focus on the underlying parameter λ and consider tuning a CUSUM to shift from the in-control mean of λ_0 to $\lambda_1 > \lambda_0$. The reference value of the CUSUM is given by

$$\begin{aligned} k^+ &= \frac{W_n\lambda_1 - W_n\lambda_0}{\ln[W_n\lambda_1] - \ln[W_n\lambda_0]} \\ &= W_n\frac{\lambda_1 - \lambda^0}{\ln\lambda_1 - \ln\lambda_0} \\ &= W_n k_u^+. \end{aligned}$$

As this shows, the reference value is the sample size W_n multiplied by k_u^+, a standard reference value based on just the *per unit* rate of the Poisson process.

A weighted Poisson CUSUM can then be defined by

$$\begin{aligned} C_0^+ &= H^+ \\ C_n^+ &= \max(0, C_{n-1}^+ + X_n - W_n k_u^+) \end{aligned}$$

Plotting C_n^+ against n provides a CUSUM scheme that can be used to detect changes in the rate parameter λ. This form, though, does not allow you to follow up a signal in the usual way by using the slope of the segment leading to the signal to estimate the change in λ. A CUSUM that does permit this estimation is given by plotting C_n^+ against $\sum_{j=1}^n W_j$, whose slope gives an estimate of $\lambda_{new} - k_u^+$.

5.6 The binomial distribution

The second common statistical model arising with discrete data is the binomial distribution. It is a natural model for the situation in which each item measured can be classified into one of two classes, traditionally but not exclusively those that "conform" to some specification and those that "do not conform" to the specification.

5.6.1 Background

The number of nonconforming items in a sample of size m follows a binomial distribution with probability π if the following four conditions hold.

1. The sample size m is fixed in advance;
2. Each item is either conforming or nonconforming;
3. The probability that an item is nonconforming is π, a constant;
4. The different items are statistically independent of each other.

If the number of nonconforming items is binomial with a probability π, then the number of conforming items will be binomial with a probability $1-\pi$, so from the point of view of distribution, it is irrelevant and a matter of convenience whether we count conforming or nonconforming items.

The binomial distribution can fail to apply for many reasons. Matching the four requirements, it may fail to hold if the sample size m is not fixed in advance. For example, we might decide to continue sampling until we had some predetermined number of nonconforming items. This means that m is not fixed in advance, and the binomial distribution does not apply. (This type of sampling is called "inverse sampling" and is very effective for monitoring situations in which π is small. It should not be overlooked just because it does not give binomial data; just be sure to implement it using the correct sampling distribution, which is negative binomial).

More commonly, the binomial distribution fails to hold because of violations in the third or fourth of its model requirements. For example, if a process is not in statistical control, then the proportion of nonconforming production vary over time. If we then assemble a sample taken over some period in which π was varying, then the number of nonconforming items in the sample will not follow a binomial distribution.

Independence can fail when a single cause can produce a cluster of adjacent nonconforming items. If this happens, then again the binomial distribution not hold.

Both the clusters and the slowly varying π situations will lead to overdispersed data in which there is more variability than the binomial distribution would predict.

The form of the binomial distribution is

$$f(x) = \frac{m!}{x!(m-x)!}\pi^x(1-\pi)^{m-x}$$

This distribution is a member of the exponential family, and the optimal CUSUM is found in Equations 6.27 through 6.29 (see Chapter 6). For upward shifts, we have:

$$\begin{aligned} C_0^+ &= H^+ \\ C_n^+ &= \max(0, C_{n-1}^+ + X_n - mk_u^+) \end{aligned}$$

and for downward shifts:

$$\begin{aligned} C_0^- &= -H^- \\ C_n^- &= \min(0, C_{n-1}^- + X_n - mk_u^-), \end{aligned}$$

where the upward and downward reference values k^+ and k^- are both of the form mk_u with k_u a *per unit* reference value dependent only on the in-control probability π_0 and the out-of-control probability π_1, and are defined by

$$k_u = -\frac{\ln\left(\frac{1-\pi_1}{1-\pi_0}\right)}{\ln\left(\frac{\pi_1(1-\pi_0)}{\pi_0(1-\pi_1)}\right)}.$$

Thus for a given in-control and out-of-control pair of π values, the basic unit reference value k_u is scaled up or down in direct proportion to the sample size.

5.6.2 Examples

Suppose that a process while in control has a proportion of nonconforming items of $\pi_0 = 0.01$. We design a CUSUM for an upward shift to $\pi_1 = 0.03$. The unit reference value for this CUSUM is

$$\begin{aligned} k_u^+ &= -\frac{\ln \frac{0.97}{0.99}}{\ln \frac{0.03\times 0.99}{0.01\times 0.97}} \\ &= -\frac{-0.0204}{1.119} \\ &= 0.018. \end{aligned}$$

A downward shift CUSUM might sensibly choose the target $\pi_1 = 0.005$. For this choice, the unit reference value on the downward CUSUM would be

$$\begin{aligned} k_u^- &= \frac{\ln \frac{0.995}{0.99}}{\ln \frac{0.005\times 0.99}{0.01\times 0.995}} \\ &= -\frac{0.00504}{-0.6982} \\ &= 0.0072. \end{aligned}$$

Note again that the reference values do not lie exactly halfway between the in-control and out-of-control parameter values in either of these CUSUMs, although both values are close.

Thus the combined CUSUM scheme for this problem would be:

$$\begin{aligned} C_0^+ &= H^+ \\ C_0^- &= -H^- \\ C_n^+ &= \max(0, C_{n-1}^+ + X_n - 0.018m) \\ C_n^- &= \min(0, C_{n-1}^- + X_n - 0.0072m). \end{aligned}$$

Suppose the sampling scheme consisted of taking rational groups of size $m = 100$. Then the schemes would evaluate to:

$$C_n^+ = \max(0, C_{n-1}^+ + X_n - 1.8)$$

$$C_n^- = \min(0, C_{n-1}^- + X_n - 0.72).$$

Following the general practice as with Poisson CUSUMs, we may round these reference values. This rounding speed up the ARL calculations and also simplify the arithmetic of updating the CUSUM if this is to be done by hand. The suggested rounding for this example is to leave k^+ alone (as it has only one decimal), but to round k^- to 0.7.

Taking a step back, with $m = 100$, while the process is in control we expect an average of 1 nonconforming item per sample. The CUSUM is designed for an upward shift to an average of 3 per sample or a downward shift of an average of 1 every other sample. These are small numbers, and it would be a major triumph of any control scheme to be able to detect shifts this small in a reasonable time.

Aiming for an in-control ARL of 1,000 for each of these CUSUMs and running the setup through ANYGETH gives for the upward CUSUM

```
 Enter the sample size and success probability
 100 .01

What is the Winsorizing constant?
1000

Do you want zero-start (say Z) or FIR (say F)?
z
Enter k and ARL (zeroes stops run)
1.8 1000
k    1.8000 h  1.00000 arls      12.6      12.5
k    1.8000 h  2.00000 arls      42.9      39.9
k    1.8000 h  3.00000 arls     142.1     136.9
k    1.8000 h  5.00000 arls    1354.3    1329.3
k    1.8000 h  4.00000 arls     447.9     432.1
k    1.8000 h  4.60000 arls     806.2     786.4
k    1.8000 h  4.80000 arls    1072.4    1044.8

k  1.8000 h  4.6000 ARL    806.23
          h  4.8000 ARL   1072.35
```

and the downward CUSUM gives

```
k     .7000 h  1.00000 arls       8.2       6.1
k     .7000 h  2.00000 arls      24.3      20.0
k     .7000 h  3.00000 arls      71.6      61.2
k     .7000 h  4.00000 arls     163.7     144.4
k     .7000 h  5.00000 arls     370.2     336.6
```

```
k     .7000 h  6.00000 arls     820.7     769.1
k     .7000 h  7.00000 arls    1813.4    1727.3
k     .7000 h  6.50000 arls    1228.8    1164.3
k     .7000 h  6.30000 arls    1047.7     987.6
k     .7000 h  6.20000 arls     973.1     914.8

k   .7000 h  6.2000 ARL     973.13
          h  6.3000 ARL    1047.71
```

The choices $h^+ = 4.8$ for the upward CUSUM and $h^- = 6.3$ for the downward CUSUM give very close to the target in-control ARL of 1,000 for each of the 2 parts of the CUSUM, and so the combined CUSUM have an ARL slightly over 500.

The performance of the CUSUM following actual shifts can be figured out using a few runs of ANYARL. This gives

```
Enter n and p
 100    .01500
 k=  1.8000  h=  4.8000 regular ARL     58.03 FIR ARL     49.88
 100    .02000
 k=  1.8000  h=  4.8000 regular ARL     14.50 FIR ARL     10.52
 100    .02500
 k=  1.8000  h=  4.8000 regular ARL      7.26 FIR ARL      4.79
```

The response to these increases is quite fast. With rational groups of size 100, the average total number of nonconforming items that would be seen from the time of shift to the signal would be 87 versus an expected 58 for the first shift; 29 versus an expected 14.5 for the second shift, and 18 versus an expected 7 for the third shift. Given the large random variability inherent in binomial data, to detect the shift and give an estimate of its time of occurrence with modest information such as this is quite impressive.

Turning to the downward shift, some ANYARL runs give

```
   Binomial down
  Enter the sample size and success probability
   100     .009
 k=    .7000  h=  6.3000 regular ARL    378.53 FIR ARL    338.40
   100     .008
 k=    .7000  h=  6.3000 regular ARL    154.04 FIR ARL    126.41
   100     .007
 k=    .7000  h=  6.3000 regular ARL     73.85 FIR ARL     54.16
   100     .006
 k=    .7000  h=  6.3000 regular ARL     42.13 FIR ARL     27.57
   100     .005
 k=    .7000  h=  6.3000 regular ARL     27.76 FIR ARL     16.59
```

These performance figures are also surprisingly good. Decreases in π are harder to detect and diagnose than increases, as intuition suggests and as these figures confirm.

There is a lower limit on the possible ARL of the downward CUSUM starting out from zero. With $k^- = 0.7$ and $h^- = 6.3$, it would take 9 successive zero readings for the downward CUSUM to cross its decision interval, starting out from zero. This holds even if π were to drop to zero so that nonconforming items stopped appearing entirely; if the CUSUM started out from zero it would still take 9 successive zero readings before the CUSUM signaled. Or more generally, the downward CUSUM from a zero start must have an ARL of at least h^-/k^-, which in this case is 9.

5.6.3 The choice of m

The size of rational group m is not particularly important in the CUSUM chart. Say we decided it was more convenient to use rational groups of size $m = 50$ than of size $m = 100$. Then the reference value for the upward CUSUM would become $k^+ = 0.018 \times 50 = 0.9$. When we had 100 observations per sample, we used an in-control ARL of 1,000; if we decrease the size of the rational group to 50, then an in-control ARL of 2,000 would yield the same total number of sampled items before a false alarm, and this seems to be a reasonable common denominator.

Designing for an in-control ARL of 2,000 using ANYGETH then gives $h^+ = 5.2$. Running the out-of-control parameter value at $\pi = 0.015$ then gives ARL $= 116.3$, against 58.3 for $m = 100$. It takes about twice as many of the samples of size 50 against samples of size 100 to detect the change, but the total number of sampled items is about the same. Counting actual sampled items, the scheme with $m = 50$ looks at an average of $5815 = 116.3 \cdot 50$, whereas the scheme with $m = 100$ looks at an average of $5830 = 58.3 \cdot 100$; these numbers are within roundoff error of each other illustrating, once again, that there is generally nothing except some convenience to be gained by using rational groups, and little to be lost by using rational groups of reasonable size.

Similar results hold for the other shifts. For example, at $\pi = 0.025$ the $m = 50$ scheme has an ARL of 14.3, or 715 sampled items, and the $m = 100$ scheme has an ARL of 7.26, or 726 sampled items.

When looking at CUSUMs for normal means, we saw that it was almost irrelevant whether the CUSUM scheme was applied to individual observations or to rational group means except for very large shifts, where the individual schemes became better. This result showing the insensitivity of the binomial scheme to the exact choice of sample size m is essentially the same result. It shows that the sample size can be chosen on grounds of convenience of sampling, and does not have to be driven by performance concerns.

We could in principle take this point right to the limit, and plot each individual sampled unit on the CUSUM. From the purely statistical viewpoint, this would be perfectly fine (just as it was in the CUSUM for a normal mean), but it is not common practice. The reason is that binomial observations tend to be much cheaper than actual measurements, and so the effort of maintaining the charts starts to become a much more significant piece of the total work. The grouping reduces the clerical charting effort.

Choose m for your convenience.

5.7 Weighted binomial CUSUMs

So far we have discussed only the case where the same sample size is taken on all occasions. This is generally advisable on both statistical and administrative grounds, but in some situations it may be necessary to have a variable sample size. This leads to a weighted CUSUM scheme of the same general form as we have seen in previous sections.

Let m_n be the sample size drawn on the nth sampling time. Then the weighted CUSUM for an upward shift is defined by

$$C_n^+ = \max(0, C_{n-1}^+ + X_n - m_n k_u^+);$$

and the downward CUSUM is defined similarly

$$C_n^- = \min(0, C_{n-1}^- + X_n - m_n k_u^-).$$

This CUSUM may be plotted against n, or it may be plotted against $\sum_1^n m_j$. Both CUSUMs signal if C_n^+ crosses the same decision interval h^+, but the advantage of plotting against the cumulative total m_n rather than n is that it gives an immediate graphic estimate of the magnitude of the shift in π.

5.7.1 Example

We illustrate the weighted CUSUM with some data on sentences in a reference book by Person (1996) - "Special Edition Using Windows 95." We took the first roughly 100 sentences in each of three chapters, and counted m_n (the total number of words in the nth sentence), and X_n (the number of long words — words of three syllables or more). The 192 sampled sentences from the first two chapters had a total of 3,291 words, 434 of them long, giving an average sentence length of 17.1 words and a proportion of long words of 0.132. We use these values to calibrate a CUSUM to check for changes in the proportion of long words.

The in-control π is 0.132, giving 2.24 long words in the average sentence. We set the reference values at $k_u^+ = 0.147$, corresponding to 2.5 long words per sentence, and $k_u^- = 0.118$, corresponding to 2.0 long words per sentence. An in-control ARL of 1,000 words would correspond to approximately $1000/17 = 59$ sentences, and the decision intervals that would give this false alarm rate are given by ANYGETH as:

```
Which distribution do you want?
   Binomial up
  Enter the sample size and success probability
   17     .132
 What are the Winsorizing constants?
   -1000.00      1000.00
 Do you want zero-start (say Z) or FIR (say F)?
 z
 Enter k and h values (zeroes stops run)
      2.5000           59.0
 k    2.5000 h  8.00000 arls      110.6       94.2
 k    2.5000 h  7.00000 arls       78.7       66.2
 k    2.5000 h  6.50000 arls       65.9       56.7

 k  2.5000 h  6.0000 ARL       54.68
           h  6.5000 ARL       65.86


 Which distribution do you want?
   Binomial down
  Enter the sample size and success probability
   17     .132
 What are the Winsorizing constants?
   -1000.00      1000.00
 Do you want zero-start (say Z) or FIR (say F)?
 z
 Enter k and ARL values (zeroes stops run)
      2.0000           59.0
 k    2.0000 h  8.00000 arls      124.1      102.6
 k    2.0000 h  7.00000 arls       86.6       74.5

 k  2.0000 h  6.0000 ARL       58.80
           h  7.0000 ARL       86.57
```

So decision intervals of $h^+ = 6$ and $h^- = 6$ would give very close to the requested ARL.

Putting the first few sentences of the third chapter through this CUSUM gives Figure 5.2. It has not taken long for the CUSUM to warn that the third chapter has a much higher proportion of long words than the first two: the CUSUM crosses the decision interval at the ninth sentence and continues rising steadily. If we carry on to the end of the sampled sentences, the average proportion of long words turns out to be 0.189.

This is actually not a very good application of a CUSUM. CUSUMs are intended for the situation where the change in the parameter occurs

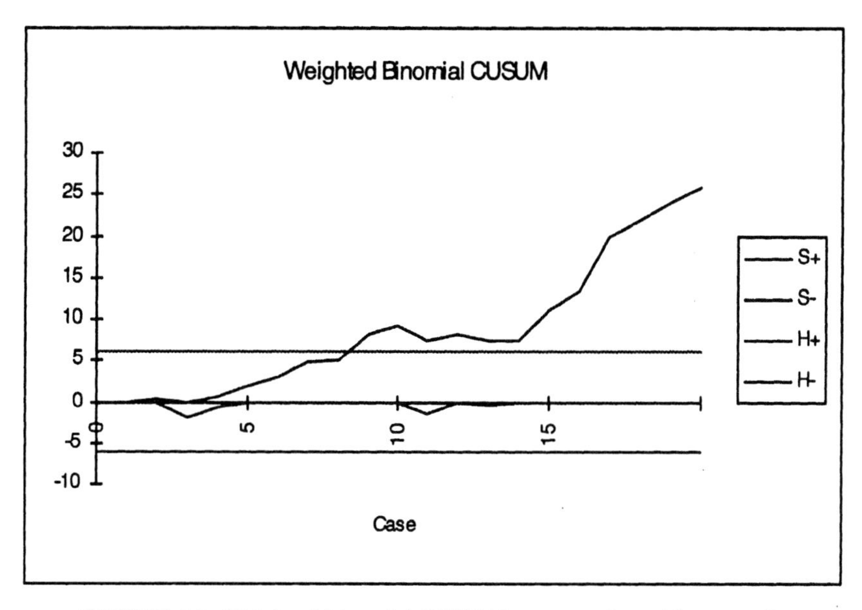

FIGURE 5.2. Weighted binomial CUSUM - proportion of long words.

at some unknown arbitrary instant. Since the chapters of this book were written by different authors, there is every reason to expect that changes in style markers such as the fraction of long words would happen when going from one chapter to the next, so the key feature of CUSUMs that makes them valuable in most uses — that they diagnose the presence of shifts occurring at unknown and unexpected times — is not present here. This is not to say that this same technology would not work well in quite similar bibliometric settings.

5.8 Other discrete distributions

We briefly mentioned the negative binomial distribution. This distribution arises in several ways, three of which are relevant to SPC problems:

- in a Poisson process in which the rate constant λ varies randomly, and follows a gamma distribution;

- in a Poisson process in which each "event" creates, not one, but a number of counted "incidents." If the number of these counted incidents per event follows a very specific distribution (the log series distribution) the total number of incidents then follow a negative binomial distribution;

- when items are characterized as either conforming or nonconforming we can either fix the total sample size and count the number of nonconforming items found (this is called "direct sampling"), or we can fix some number of nonconforming items, carry on sampling until we have found that target number of nonconforming items, and then use as our process measure the number of conforming items seen in the sample (this is called "inverse sampling.") In inverse sampling, the number of conforming items seen follows a negative binomial distribution.

In the first two situations, the negative binomial distribution is a way to enrich the basic Poisson scheme to account for departures from the ideal of the Poisson process. The value of this sort of enrichment was shown when we looked at the in-control ARL that resulted when a process was modeled using Poisson calculations, but actually followed a negative binomial distribution with even quite small overdispersion.

The third situation provides a good mechanism for handling sampling problems where π is small. The problem with direct sampling is that if π is small, fixed-size samples may give no nonconforming items, so that the information they give on π is quite limited. Inverse sampling, by ensuring that nonconforming items be seen, tends to provide a more reliable level of information.

The negative binomial distribution has probability distribution

$$f(x) = \frac{\Gamma(r+x)c^r}{x!\Gamma(r)(1+c)^{r+x}}, \quad x = 0, 1, \ldots$$

and its mean is given by $\mu = r/c$. If we take r as given and concentrate on monitoring sequences of data for a change in c from an in-control level c_0 to an out-of-control level $c_1, c_1 > c_0$, the decision interval CUSUM is given in Equations 6.30 and 6.31, with

$$\begin{aligned} C_0 &= H^+ \\ C_n^+ &= \max(0, C_{n-1} + X_n - k^+) \\ k^+ &= \frac{r \ln \frac{c_0(1+c_1)}{c_1(1+c_0)}}{\ln(\frac{1+c_0}{1+c_1})} \end{aligned}$$

and the CUSUM for a downward shift in c uses

$$k^- = \frac{r \ln \frac{c_1(1+c_0)}{c_0(1+c_1)}}{\ln(\frac{1+c_0}{1+c_1})}.$$

As an example of use of the negative binomial CUSUM, we again use the sentence lengths of the Windows 95 reference book. The sentence lengths

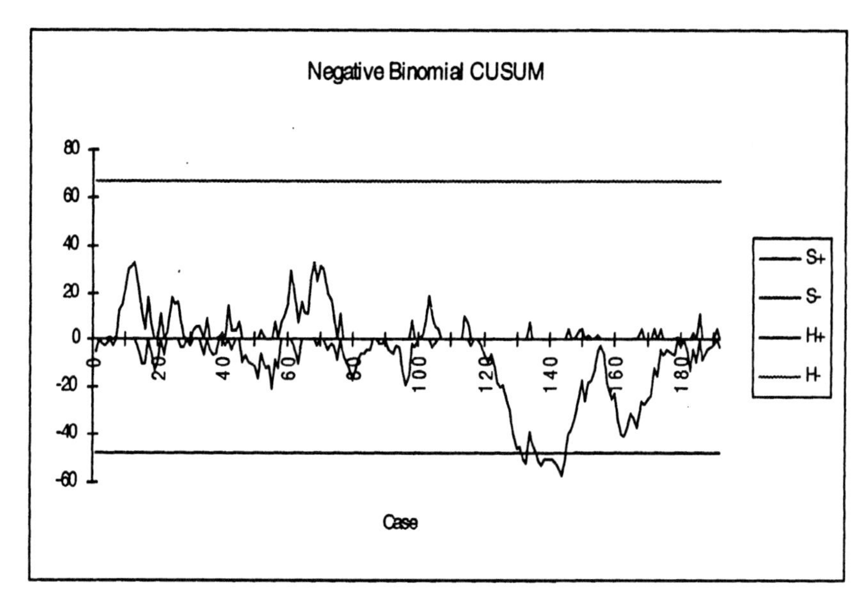

FIGURE 5.3. Sentence length data decision interval CUSUMs.

from the first chapter averaged 18.02 words, with a variance of 62.16. Clearly these numbers are nowhere near Poisson distributed.

The sentence lengths seem to be modeled reasonably well by a negative binomial with $r = 7.36$ and $c = 0.41$. (We obtained these values of r and c by using Equations 5.1 and 5.2.) We'll use these values to set up a CUSUM to monitor the sentence length elsewhere in the book. Setting the upward reference value at $k^+ = 21$ and the downward value at $k^- = 15$, the ANYGETH shows that the decision interval

```
   Neg binomial up

Give the mean and variance (variance > mean)
    18.01       62.16

What are the Winsorizing constants?
   -1000.00      1000.00

Do you want zero-start (say Z) or FIR (say F)?
z

Enter k and ARL values (zeroes stops run)
    21.0000       2000.0

k  21.0000 h 65.00000 arls   1751.1    1690.5
k  21.0000 h 68.00000 arls   2237.1    2163.8
k  21.0000 h 67.00000 arls   2061.9    1995.2
```

```
 k  21.0000 h 66.00000 arls    1900.2    1833.5
 k 21.0000 h 66.0000 ARL    1900.18
           h 67.0000 ARL    2061.85

   Neg binomial down

 Give the mean and variance (variance > mean)
 18.02      62.16

What are the Winsorizing constants?
   -1000.00      1000.00

Do you want zero-start (say Z) or FIR (say F)?
 z

Enter k and ARL values (zeroes stops run)
     15.0000        2000.0

 k  15.0000 h 45.00000 arls    1451.7    1395.2
 k  15.0000 h 48.00000 arls    2085.4    2011.1
 k  15.0000 h 47.00000 arls    1848.4    1783.6
 k 15.0000 h 47.0000 ARL    1848.41
           h 48.0000 ARL    2085.36
```

In-control ARLs of approximately 2,000 are given by $h^+ = 67$ and $h^- = 48$. We use these values to check the sentence length starting from the second chapter and running on into the third. This gives Figure 5.3.

The CUSUM shows a sharp downward move about 30 sentences into the third chapter, following which the C^- CUSUM moves back up suggesting that the shortened sentences were temporary.

5.9 Summary

In this chapter, we discussed the two most common models for integer data, the Poisson model and the binomial model. We illustrated the design of CUSUMs for both models, using software to find the decision intervals. We paid attention to the effects of model departures, and showed how to construct weighted CUSUMs.

Discrete data poses special problems for CUSUM methods, since ARLs also assume discrete values. This often forces us to accept ARLs that are only approximately what we want.

5.10 Further reading

For problems involving 100% inspection and go/no-go measures, Bourke (1992) proposes a CUSUM of the gaps between successive nonconforming items. These gaps follow a geometric distribution, a special case of the negative binomial

More generally, there is a close connection between several of the CUSUM problems we've been discussing.

- If events occur according to a Poisson process, then the total number of events occurring in a fixed time interval follows a Poisson distribution. The time required for r events to occur follows a gamma distribution. Thus a chart in which we had total elapsed time on the horizontal axis, and total number of events on the vertical axis could be thought of as an uncentered V-mask CUSUM plot of Poisson data. But by flipping the chart on its side, it would become an uncentered weighted V-mask CUSUM of exponential data.

- Similarly, if a stream consists of individual items that are either conforming or nonconforming and the probability of conformity is a fixed value π, then the total number of conforming items in a sample of fixed size n follows a binomial distribution. The total number of items seen until m conforming items have been found is a negative binomial. A plot of the running totals of items seen and conforming items seen could then be thought of as an uncentered V-mask CUSUM of either binomial or of negative binomial data.

A "CUSUM in disguise" is outlined in Lucas (1989). This is a binomial control scheme that signals if there are at least two nonconforming items with t successive sampled items. This scheme (which is effective for low nonconforming rates) is equivalent to a binomial CUSUM with $n = 1$, $k = 1/t$, and $h = 1$.

6
Theoretical foundations of the CUSUM

> Most of the fundamental ideas of science are essentially simple, and may, as a rule, be expressed in a language comprehensible to everyone.
>
> *Albert Einstein*

This chapter gives the technical underpinnings of the rest of the book. It may be used as a reference as necessary, and skipped without loss of continuity.

6.1 General theory

There are two threads of theoretical work that converge to provide the CUSUM's technical underpinning. One is Page's proposal (Page, 1954). In this, he proposed the CUSUM of deviations from the reference value, and suggested that it be diagnosed by looking at the maximum rise from one point in the CUSUM to a subsequent point.

A different thread of work was the sequential probability ratio test (SPRT), developed by Abraham Wald (1945). Like the CUSUM, the SPRT uses data sequentially, adding the information from each new observation X_n as it becomes available.

The SPRT tests a simple null hypothesis (H_0) against a simple alternative hypothesis (H_1). Associated with each hypothesis is a probability density or mass function $f_0(x)$ and $f_1(x)$, respectively, for the measure X_n.

Assume that we have a sequence of independent observations $\{X_i\}$ of length n and we wish to decide between H_0 and H_1. In the SPRT, we form the likelihood ratio Λ_n given by

$$\Lambda_n = \prod_{i=1}^{n} \frac{f_1(X_i)}{f_0(X_i)}.$$

The test accepts H_0 if Λ_n is less than or equal to some cutoff value A, and rejects H_0 in favor of H_1 if Λ_n is bigger than another cutoff constant B. And if $A < \Lambda_n < B$, then the SPRT calls for another observation X_{n+1} and updates the likelihood ratio to incorporate it.

It is usually easier to work with logarithms of the likelihood ratio

$$\ln \Lambda_n = \sum_{i=1}^{n} \ln\left(\frac{f_1(X_i)}{f_0(X_i)}\right).$$

Notice that we can write $\ln \Lambda_n$ as

$$\begin{aligned} \ln \Lambda_n &= \sum_{i=1}^{n} Z_i; \\ Z_i &= \ln\left(\frac{f_1(X_i)}{f_0(X_i)}\right). \end{aligned}$$

The score variable Z_i is just a transformation of the random quantity X_i, and so is also a random variable whose distribution can be calculated from that of the X. Under the assumption that the original X_is are independent, these Z_is are also independent.

The SPRT then

- accepts H_0 if $\ln \Lambda_n \leq \ln A$,
- accepts H_1 if $\ln \Lambda_n \geq \ln B$, and
- calls for another observation if $\ln A < \ln \Lambda_n < \ln B$.

This is illustrated in Figure 6.1, with $\ln A = -3$ and $\ln B = 3$, where the hypothesis H_1 would have been accepted at observation number 4.

Notice the striking resemblance between the graph of the SPRT in Figure 6.1 and the original V-mask form of the CUSUM.

Wald conjectured (Wald, 1947) and later proved (Wald and Wolfowitz, 1948) that the sequential probability ratio test was optimal for deciding between two point hypotheses in the sense that the expected number of points sampled before a decision could be reached was minimized with the SPRT. A precise statement of these optimality properties of the SPRT in a decision framework can be found in Ferguson (1967) .

In practice, SPRT tests work with this log-likelihood, or $\ln(\Lambda_n)$, so that the test is based on the cumulative sum of the Z_i. We decide to accept,

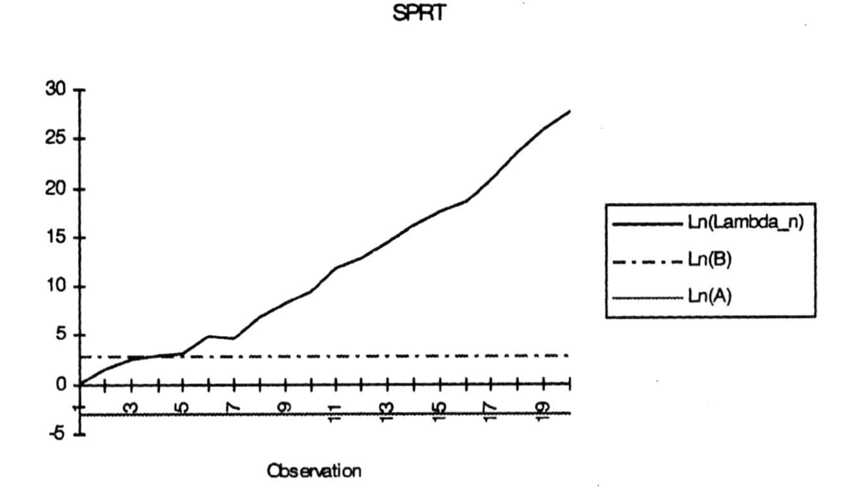

FIGURE 6.1. A graphical description of the SPRT.

reject, or continue sampling based on the value of this cumulative sum. If the hypothesis H_1 is true, the Z_is will have a positive expected value. As points are added, the cumulative sum tends to drift upward. If the hypothesis H_0 is true, the Z_is will have a negative expected value, and the sum will tend to drift downward.

The constants A and B determine the probabilities of Type I and Type II error. Deciding on the Type I and II error probabilities fixes the required values for A and B.

6.1.1 Relation of the SPRT to the CUSUM

Page (1954) explicitly recognized that his CUSUM rules resulted in a "sequence of Wald sequential tests." The one-sided CUSUM can be thought of as a repeated SPRT. The primary difference between a CUSUM and Wald's SPRT is that in the CUSUM, the **hypothesis of in-control, H_0, is never accepted.** We are never content to decide that the process is in control and stop sampling. Instead, we restart the test each time the evidence favors the hypothesis that we are in control.

The evidence favors the null (in-control) hypothesis whenever the cumulative sum of Z_i is negative. Whenever that happens, we start over by resetting the sum to zero, or, in algebraic terms,

$$C_n^+ = \max(0, C_{n-1}^+ + Z_n).$$

In the CUSUM approach, we sample until we reject the null hypothesis in favor of the alternative.

The "score" function Z_i turns out to be the same function seen in the SPRT; it is a function of X_i, and depends on any parameters that the in-control and out-of-control distributions $f_0(X_i)$ and $f_1(X_i)$ may have.

For distributions that are members of the exponential family, Z_i is a particularly simple function, as we show in the following. This development of the SPRT as a cumulative sum leads directly to many of the standard CUSUM schemes.

6.1.2 Optimality properties

Various optimality properties of the CUSUM have been shown by Lorden (1971), Moustakides (1986), Banzal and Papantoni-Kazakos (1986), Ritov (1990), and Yashchin (1993a).

How do we define optimality? Following Lorden (1971) and Moustakides (1986), we consider the problem of determining when a process has shifted from one known distribution to a second known distribution. We consider the first known distribution to be the in-control distribution, and the second distribution to be a known out-of-control state. The shift occurs at some unknown time m.

There are many schemes that can detect this change. The standard Shewhart scheme is an obvious example. We classify these schemes by their "false alarm rates", the expected time until the process signals while it remains in control. Among all procedures with the same false alarm rates, the optimal procedure is that which detects changes the quickest. Or more formally, among all procedures with the same in-control ARL, the optimal procedure has the smallest expected time until it signals a change, once the process shifts to the out-of-control state.

Moustakides (1986) proved that the CUSUM scheme was optimal in this sense. Among all tests with the same in-control ARL, the CUSUM had the smallest expected run length out-of-control. The reader is referred to Moustakides for a precise statement and proof.

The optimality of the CUSUM is for detecting the shift to a single specific out-of-control distribution. The CUSUM that is optimal for one particular change — for example, a one-standard-deviation increase in mean — is not optimal for detecting some other change — for example, a two-standard-deviation increase in mean. For detecting a two-standard-deviation shift in mean, a different CUSUM would be optimal. This means that general statements such as "the CUSUM is optimal for detecting persistent shifts" need to be qualified with the restriction that the CUSUM is optimal for detecting the specific persistent shift for which it was designed.

In practice, there usually is no single out-of-control distribution to which the process would shift; rather, there is a variety of possible shifts. In hypothesis testing terms, the alternative of real interest is not a point hypothesis but an interval hypothesis. We get around this difficulty by selecting a particular out-of-control state(s) for which we want the CUSUM to be optimal. Although this CUSUM will be the optimal diagnostic only if the process shifts to the particular alternative for which it is designed, most CUSUMs have quite robust performance. That is to say, while the CUSUM for detecting a one-standard-deviation shift is only the optimal diagnostic for that particular shift, it does nearly as well as the optimal CUSUM for all shifts "not too far" from one standard deviation. In Chapter 3, Figure 3.1 illustrated this general truth in the context of a CUSUM for a normal mean.

6.2 The general exponential family

In this section, we follow the notation of Hawkins (1992b). The exponential family of distributions is important in theoretical statistics because of its good inferential properties, and it is also important in applied statistics because it includes many useful distributions. The probability density (or mass function if discrete) for any member of the exponential family with a single parameter θ can be written as

$$f(y|\theta) = \exp\left\{(a(y)b(\theta) + c(y) + d(\theta)\right\}, \tag{6.1}$$

where θ is the parameter of the distribution and Y is the corresponding random variable. This formulation holds for both discrete and continuous members of the exponential family.

The function $a(y)$ gives the "minimal sufficient statistic." It is a distillation of all the information about θ in an observation. The function $b(\theta)$ is called the "natural parameter." The joint density for a random sample of Y, where Y is a member of the exponential family, is given by

$$f(\mathbf{y}|\theta) = \exp\left(\sum_{i=1}^{n} a(y_i)b(\theta) + \sum_{i=1}^{n} c(y_i) + nd(\theta)\right).$$

Suppose we want to test whether the process has gone from an in-control parameter value θ_0 to an out-of-control value θ_1. Then the "score" variables Z_i simplify to

$$\begin{aligned} Z_i &= \ln\left(\frac{f_1(Y_i)}{f_0(Y_i)}\right) \\ &= a(Y_i)\{b(\theta_1) - b(\theta_0)\} + \{d(\theta_1) - d(\theta_0)\}. \end{aligned}$$

A generic decision interval CUSUM with its recursion

$$D_n = \max(0, D_{n-1} + Z_n)$$

with the signal if $D_n > A$ then specializes to

$$D_n = \max\{0, D_{n-1} + a(Y_i)\{b(\theta_1) - b(\theta_0)\} + d(\theta_1) - d(\theta_0)\} \tag{6.2}$$

signaling if $D_n > A$.

We write $X_n = a(Y_n)$ for the sufficient statistic, and define

$$k = -\frac{d(\theta_1) - d(\theta_0)}{b(\theta_1) - b(\theta_0)}. \tag{6.3}$$

If $b(\theta_1) - b(\theta_0) > 0$, then we can rescale by dividing the CUSUM in Equation 6.2 through by $(b(\theta_1) - b(\theta_0))$, getting the recursion

$$C_n^+ = \max(0, C_{n-1}^+ + X_n - k), \tag{6.4}$$

where $C_n^+ = D_n/(b(\theta_1) - b(\theta_0))$ and the CUSUM signals if $C_n^+ > h^+$, where $h^+ = A/(b(\theta_1) - b(\theta_0))$.

If $(b(\theta_1) - b(\theta_0)) < 0$, then dividing the CUSUM through by $b(\theta_1) - b(\theta_0)$ reverses the direction of the CUSUM, giving the recursion

$$C_n^- = \min(0, C_{n-1}^- + X_n - k), \tag{6.5}$$

where $C_n^- = D_n/(b(\theta_1) - b(\theta_0))$ and the CUSUM signals if $C_n^- < -h^-$, where $h^- = A/(b(\theta_1) - b(\theta_0))$.

In this chapter, we start our CUSUMs at zero. We note that one can choose other head starts as discussed in Chapter 3. Using a head start does not affect the choice of $a(y)$ or k. Head starts do affect average run lengths in and out of control, which in turn affects the selection of h to achieve a given ARL.

> Notice that k is completely determined by the choice of the two parameter values associated with H_0 and H_1.

The CUSUM differs from the SPRT: we never accept the null hypothesis that we are in control. This means our probability of a Type II error is zero: we never accept the null hypothesis when it is false.

Consider the decision interval form of the CUSUM. Any time the CUSUM is negative, which favors the in-control null hypothesis, we start over by resetting the cumulative sum to zero.

The h^+ and h^- parameters are still to be specified. Since the CUSUM gives the smallest out-of-control ARL for all tests with the same ARL in control, we set h by specifying the desired ARL in-control. This requires us to have means for finding the ARL in control for a given h, discussed later in this chapter.

> For members of the single parameter exponential family, the optimal CUSUM design is completely specified by the selection of the in-control parameter, the out-of-control parameter, and the ARL in control.

6.2.1 Derivation of the CUSUM for a normal mean shift

Given the general formulation for the CUSUM expressed in Equation 6.3, we can easily derive the CUSUM for an upward or downward shift in the mean of a normal distribution, with the variance known.

Let σ be fixed and known. We can write the normal density in "exponential family" form as

$$f(y|\mu) = \exp\left(\frac{y\mu}{\sigma^2} - \frac{y^2}{2\sigma^2} - \ln(\sigma\sqrt{2\pi}) - \frac{\mu^2}{2\sigma^2}\right)$$

from which it follows that

$$\begin{aligned} a(y) &= y \\ b(\mu) &= \frac{\mu}{\sigma^2} \\ d(\mu) &= -\frac{\mu^2}{2\sigma^2} \\ k &= -\frac{(-\mu_1^2 + \mu_0^2)/(2\sigma^2)}{(\mu_1 - \mu_0)/\sigma^2} \\ &= \frac{\mu_0 + \mu_1}{2}. \end{aligned}$$

Applying Equations 6.3 and 6.4, the following CUSUM scheme for the mean μ_0 of a normal distribution, tuned to the alternate value $\mu_1 > \mu_0$, results

$$\begin{aligned} C_0^+ &= 0 \\ C_{n+1}^+ &= \max(0, C_n^+ + Y_n - k^+) \\ k^+ &= \frac{\mu_0 + \mu_1}{2}. \end{aligned}$$

This scheme signals if $C_n^+ > h^+$, where h^+ is chosen so the scheme has a specified ARL in control.

If $\mu_1 < \mu_0$, then the CUSUM looks very much the same, except that its direction has changed because of the division of the fundamental standard form by the negative $b(\theta_1) - b(\theta_0)$. The CUSUM is then

$$\begin{aligned} C_0^- &= 0 \\ C_{n+1}^- &= \min(0, C_n^- + Y_n - k^-) \\ k^- &= \frac{\mu_0 + \mu_1}{2} \end{aligned}$$

and signals if $C_n^- < -h^-$.

If we regard μ as fixed and think about monitoring for possible changes in σ, then we want to rewrite the normal density in the form

$$f(y|\sigma) = \exp\left(\frac{-(y-\mu)^2}{2\sigma^2} - \ln(\sigma) - \ln(\sqrt{2\pi})\right).$$

This identifies the components

$$\begin{aligned} a(y) &= (y-\mu)^2 \\ b(\sigma) &= -(2\sigma^2)^{-1} \\ d(\sigma) &= -\ln\sigma \end{aligned}$$

and leads to the CUSUM with reference value

$$\begin{aligned} k &= -\frac{-\ln\sigma_1 + \ln\sigma_0}{(-2\sigma_1^2)^{-1} + (2\sigma_0^2)^{-1}} \\ &= -\frac{\ln\sigma_1 - \ln\sigma_0}{(2\sigma_1^2)^{-1} - (2\sigma_0^2)^{-1}} \end{aligned}$$

and summand $X_n = (Y_n - \mu)^2$. This is the optimal diagnostic for a step change in σ.

Despite its optimality, this diagnostic is not widely used. The reason is that its summand involves μ, and so is affected by changes in μ as well as by changes in σ. We do not think that is a particularly good reason to pass up the possibility of using an optimal CUSUM. As we monitor μ separately, it should be easy to recognize the false signals on the scale chart caused by shifts in μ. In any event, we already have to live with the fact that changes in σ affect the ARL of the chart for μ and so may lead to false signals on the μ chart; it is hard to see why the mirror image potential confusion is unacceptable.

6.2.2 *The gamma family and normal variance*

The gamma distribution, $X \sim \Gamma(\alpha, \beta)$, is a member of the exponential family, and can be written with density

$$f(y|\alpha,\beta) = \exp\left(-y/\beta + (\alpha-1)\ln y - \alpha\ln\beta - \ln(\Gamma\alpha)\right). \tag{6.6}$$

The gamma distribution has two parameters. As either of them could change, we want to set up separate control schemes for changes in α and changes in β. For a CUSUM for a change in α, we regard β as fixed, and make α the single parameter of the exponential-family distribution. This gives

$$\begin{aligned} a(y) &= \ln y \\ b(\alpha) &= \alpha - 1 \\ d(\alpha) &= -(\alpha\ln\beta - \ln\Gamma(\alpha)) \end{aligned} \tag{6.7}$$

To monitor for a change in α from its in-control level α_0 to an out-of-control level α_1, we use Equation 6.3 to set up a CUSUM with reference value

$$
\begin{aligned}
k &= -\frac{(-\alpha_1 + \alpha_0)\ln\beta - \ln\Gamma(\alpha_1) + \ln\Gamma(\alpha_0)}{(\alpha_1 - 1) - (\alpha_0 - 1)} \\
&= \frac{\ln\beta(\alpha_0 - \alpha_1) - \ln\left(\frac{\Gamma(\alpha_0)}{\Gamma(\alpha_1)}\right)}{\alpha_1 - \alpha_0}. \qquad (6.8)
\end{aligned}
$$

If $\alpha_1 > \alpha_0$, then the CUSUM tests for upward shifts

$$
\begin{aligned}
C_0^+ &= 0 \\
C_n^+ &= \max(0, C_{n-1}^+ + \ln(Y_n) - k) \qquad (6.9)
\end{aligned}
$$

and signals on large positive values $C_n^+ > h^+$. If $\alpha_1 < \alpha_0$, then the CUSUM tests for downward shifts:

$$
\begin{aligned}
C_0^- &= 0 \qquad (6.10) \\
C_n^- &= \min(0, C_{n-1}^- + \ln(Y_n) - k) \qquad (6.11)
\end{aligned}
$$

and signals on large negative values $C_n^- < -h^-$.

To monitor for a change in β, we regard α as fixed and make β the single parameter of the exponential family. We have then that

$$
\begin{aligned}
a(y) &= y \\
b(\beta) &= -1/\beta \\
d(\beta) &= -\alpha\ln\beta
\end{aligned}
$$

This CUSUM has reference value

$$
k = \alpha\frac{\ln(\beta_1) - \ln(\beta_0)}{\beta_1^{-1} - \beta_0^{-1}}. \qquad (6.12)
$$

The upward CUSUM for a change in β follows as

$$
\begin{aligned}
C_0^+ &= 0 \qquad (6.13) \\
C_n^+ &= \max(0, C_{n-1}^+ + Y_n - k). \qquad (6.14)
\end{aligned}
$$

The CUSUM for a downward shift in β is

$$
\begin{aligned}
C_0^- &= 0 \qquad (6.15) \\
C_n^- &= \min(0, C_{n-1}^- + Y_n - k), \qquad (6.16)
\end{aligned}
$$

where the equation for k has the same form as for upward CUSUMs.

CUSUMs for a change in α were developed by Regula (1976) in his doctoral dissertation. Explicit CUSUMs for a change in β were developed by Olwell (1996); they also follow immediately from Johnson and Leone (1962b).

6.2.3 *Relation to normal variances*

We outlined the optimal CUSUM for a normal variance using individual observations in the previous section. If we sample rational groups of size $m > 1$, then we have another way to monitor σ, and one that does not have the objection of potentially signaling as a result of changes in μ.

Writing the sample variance for the nth sample of size m,

$$S_n^2 = \frac{\sum_{i=1}^{m}(X - \overline{X})^2}{m-1},$$

then under the normal model $X \sim N(\mu, \sigma^2)$, the variance follows a scaled chi-squared distribution:

$$\frac{(m-1)S^2}{\sigma^2} \sim \chi^2_{m-1}. \tag{6.17}$$

The chi-squared distribution is a member of the gamma family; with our parameterization of the gamma family, a chi-squared random variable $\chi^2_n \sim \Gamma(n/2, 2)$.

This means that

$$\frac{(m-1)S^2}{\sigma^2} \sim \Gamma\left(\frac{m-1}{2}, 2\right).$$

We need a few intermediate steps before we arrive at a useful CUSUM. First a distributional result on rescaling a gamma variate: if $Y \sim \Gamma(\alpha, \beta)$ and $Z = aY$, then $Z \sim \Gamma(\alpha, a\beta)$.

This means that the sample variances S_n^2 follow the gamma distribution

$$S_n^2 \sim \Gamma\left(\frac{m-1}{2}, \frac{2\sigma^2}{m-1}\right).$$

Suppose that the in-control variance is σ_0^2, and that we want to tune the CUSUM for a shift to a larger variance σ_1^2. Then substituting

$$\begin{aligned} \alpha &= \frac{m-1}{2} \\ \beta_0 &= \frac{2\sigma_0^2}{m-1} \\ \beta_1 &= \frac{2\sigma_1^2}{m-1} \end{aligned}$$

in the general formula for the reference value gives (after a little simplification)

$$k = -\alpha \frac{\ln(\beta_1) - \ln(\beta_0)}{\beta_1^{-1} - \beta_0^{-1}} \tag{6.18}$$

$$= \frac{\ln(\sigma_0^2/\sigma_1^2)\sigma_0^2\sigma_1^2}{\sigma_0^2 - \sigma_1^2}. \tag{6.19}$$

This resulting CUSUM is optimal for changes in σ^2 for samples of size greater than 1. This scheme was discussed in the V-mask form by Johnson and Kotz (1969), who derived it directly from the SPRT. If we have individual observations and not rational groups, then obviously we have no rational group variances S_n^2 and so cannot apply this CUSUM. We can, though, use the optimal CUSUM for individual observations from the previous section.

This CUSUM using the sample variances indirectly addresses another issue: whether CUSUM control should use the rational groups' *variances* or their *standard deviations* to monitor σ. The answer is that we use the variances to get the optimal CUSUM; making a CUSUM of standard deviations would not be making use of the most effective form of the information on spread.

We earlier recommended CUSUMs of individual observations. We derive the optimal CUSUM for that case now. We standardize the observations in accordance with the in-control parameters:

$$U_n = \frac{X - \mu_0}{\sigma_0}.$$

In control, $U_n^2 \sim \chi_1^2$. Assume we experience a scale change, so that the variance becomes σ_1^2. Then the distribution of U_n^2 shifts to

$$U_n^2 \sim \Gamma\left(\frac{1}{2}, \frac{2\sigma_1^2}{\sigma_0^2}\right). \tag{6.20}$$

Accordingly, the CUSUM scheme follows from Equations 6.12 and 6.19. For an upward shift:

$$k^+ = -\frac{\ln(\sigma_0/\sigma_1)}{1 - \sigma_0^2/\sigma_1^2} \tag{6.21}$$

$$S_0^+ = 0 \tag{6.22}$$

$$S_n^+ = \max(0, S_{n-1} + U_n^2 - k^+). \tag{6.23}$$

This scheme is an extension of the preceding scheme for sample variances.

6.2.4 The Poisson family

The Poisson family is also a member of the exponential family. Its probability mass function can be written in the standard form as

$$f(y|\lambda) = \exp(y \ln \lambda - \lambda - \ln \Gamma(y+1))$$

from which it follows that

$$a(y) = y$$

$$\begin{aligned} b(\lambda) &= \ln \lambda \\ d(\lambda) &= -\lambda \\ k &= \frac{\lambda_1 - \lambda_0}{\ln(\lambda_1/\lambda_0)}. \end{aligned} \tag{6.24}$$

Accordingly, the optimal CUSUM for testing a shift in λ from the in-control level λ_0 to some larger value λ_1 is given by

$$\begin{aligned} C_0^+ &= 0 \\ C_n^+ &= \max(0, C_{n-1}^+ + Y_n - k) \end{aligned} \tag{6.25}$$

with a signal if $C_n^+ \geq h^+$.

The CUSUM for a change from λ_0 to a smaller value λ_1 is of the same form, but directed downward:

$$\begin{aligned} C_0^- &= -H^- \\ C_n^- &= \min(0, C_{n-1}^- + Y_n - k) \end{aligned} \tag{6.26}$$

with a signal if $C_n^- \leq -h^-$. Johnson and Leone (1962c) derived this scheme, in the V-mask form, from the SPRT.

6.2.5 *The binomial family*

Let $Y \sim$ Binomial(m, π). Then Y has probability mass function

$$f(y) = \frac{m!}{y!(m-y)!}\pi^x(1-\pi)^{m-x}.$$

This distribution is a member of the exponential family, as we can see by rewriting it as

$$f(y|m,\pi) = \exp\left[y \ln\left(\frac{\pi}{1-\pi}\right) + m\ln(1-\pi) + \ln m! - \ln\{y!(m-y)!\}\right].$$

This form shows that

$$\begin{aligned} a(y) &= y \\ b(\pi) &= \ln\left(\frac{\pi}{1-\pi}\right) \\ d(\pi) &= m\ln(1-\pi). \end{aligned}$$

From this and Equation 6.3, we find the reference value for a shift from an in-control value π_0 to an out-of-control level of π_1 to be

$$k = -m\frac{\ln(1-\pi_1) - \ln(1-\pi_0)}{\ln\{\pi_1/(1-\pi_1)\} - \ln\{\pi_0/(1-\pi_0)\}}.$$

This simplifies to

$$k = -m \frac{\ln\left(\frac{1-\pi_1}{1-\pi_0}\right)}{\ln\left(\frac{\pi_1(1-\pi_0)}{\pi_0(1-\pi_1)}\right)}. \tag{6.27}$$

The reference value is of the form $k = mk_u$, where k_u is a function of π_0 and π_1 only. This separates out the effects of the parameter m, the number of binomial trials, from that of the in-control and out-of-control probabilities.

If $\pi_1 > \pi_0$, then the CUSUM tests for upward shifts in Y_n, and has the recursion

$$C_n^+ = \max(0, C_{n-1}^+ + Y_n - mk_u), \tag{6.28}$$

and if $\pi_1 < \pi_0$, then we get a downward CUSUM with the almost identical recursion:

$$C_n^- = \min(0, C_{n-1}^- + Y_n - mk_u). \tag{6.29}$$

Johnson and Leone (1962c) first derived this scheme in V-mask form.

6.2.6 *The negative binomial family*

The negative binomial distribution comes in several visually different forms and several different parameterizations. Two forms are as the total number of *trials* until you get r successes in a sequence of binomial trials; and as the total number of *failures* until you get r successes in a sequence of binomial trials. The first of these random variables is just r plus the second. We choose not to use either of these forms, but rather to represent the negative binomial as a general-purpose discrete distribution that can be used to model nonnegative integer-valued data.

Our form of the negative binomial distribution has probability mass function

$$f(x|r,c) = \frac{\Gamma(r+x)c^r}{x!\Gamma(r)(1+c)^{r+x}}, \quad x = 0, 1, \ldots$$

The mean is given by $\mu = r/c$ and the variance is given by $\mu(1 + 1/c)$.

We saw in Chapter 5 that the negative binomial was a plausible model for count data with heavier tails than the Poisson. If c is very large, then the negative binomial is very close to Poisson in shape.

It is convenient to fix r and set up a scheme for detecting changes in c. These changes reflect both changes in the mean of X, and also (although less dramatically) its degree of over-dispersion. With r fixed, we can write $f(x)$ as a member of the exponential family with

$$\begin{aligned} a(x) &= x \\ b(c) &= -\ln(1+c) \\ d(c) &= r\ln\left(\frac{c}{1+c}\right). \end{aligned}$$

Our reference value k^+ is found to be

$$k^+ = -\frac{r \ln\left(\frac{c_1(1+c_0)}{c_0(1+c_1)}\right)}{\ln\left(\frac{1+c_1}{1+c_0}\right)} \tag{6.30}$$

and the CUSUM scheme for an upward shift is

$$\begin{aligned} C_0^+ &= 0 \\ C_n^+ &= \max(0, C_{n-1}^+ + X_n - k^+) \end{aligned} \tag{6.31}$$

and signals when $C_n^+ > h^+$. As usual, h^+ is found by specifying the desired in-control ARL.

Note that the mean varies inversely with c. An *upward* shift in the mean of X_n corresponds to a *downward* shift in c. A downward shift in the mean of X_n is monitored by the corresponding CUSUM for an increase in c.

6.2.7 The inverse Gaussian family

The inverse Gaussian (IG) family is not as widely used as the other distributions we have mentioned. Its density function is

$$f(y) = \left(\frac{\lambda}{2\pi y^3}\right)^{1/2} \exp\left(\frac{-\lambda(y-\mu)^2}{2\mu^2 y}\right).$$

It has the two parameters λ and μ, and is within the exponential family for each of them.

Let $Y \sim IG(\mu, \lambda)$. Following our now familiar methods, we obtain the CUSUM scheme for an upward change in μ by writing the density in exponential-family terms for μ as

$$f(y, \mu) = \exp(a(y)b(\mu) + c(x) + d(\mu)),$$

where by matching terms we identify

$$\begin{aligned} a(y) &= y \\ b(\mu) &= -\frac{\lambda}{2\mu^2} \\ d(\mu) &= \lambda/\mu. \end{aligned}$$

This shows that the sufficient statistic for cumulative summation is Y_n itself, and the reference value for the shift from an in-control level μ_0 to an out-of-control level μ_1 is:

$$k = -\frac{2\mu_0\mu_1}{\mu_0 + \mu_1}.$$

If $\mu_1 > \mu_0$, then we get the upward CUSUM

$$k^+ = -\frac{2\mu_0\mu_1}{\mu_0 + \mu_1} \quad (6.32)$$

$$C_0^+ = 0$$

$$C_n = \max(0, C_{n-1} + Y_n - k^+). \quad (6.33)$$

The CUSUM for a downward shift in μ is given by just changing the "max" to a "min."

The CUSUM for a shift in λ is slightly more complicated. Rewriting the density to emphasize its dependence on λ gives

$$f(y, \lambda) = \exp(-\lambda a(y)/2 + c(y) - \ln(\lambda^{1/2})),$$

where the sufficient statistic and the $b(.)$ and $d(.)$ functions are:

$$a(y) = \frac{(y-\mu)^2}{y\mu^2} \quad (6.34)$$

$$b(\lambda) = -\frac{\lambda}{2} \quad (6.35)$$

$$d(\lambda) = -\frac{\ln(\lambda)}{2}. \quad (6.36)$$

Straightforward application of Equations 6.17, 6.12, and 6.19 gives:

$$k = -\frac{\ln(\lambda_0/\lambda_1)}{\lambda_0 - \lambda_1} \quad (6.37)$$

$$C_0 = 0$$

$$C_n = \max(0, S_n + a(Y_n) - k). \quad (6.38)$$

A smaller value for λ corresponds to larger average values for $a(y)$, so the CUSUM for a downward shift in λ is an upward CUSUM in $a(y)$; if the out-of-control value λ_1 is less than the in-control value λ_0 we use the CUSUM

$$k^+ = -\frac{\ln(\lambda_0/\lambda_1)}{\lambda_0 - \lambda_1} \quad (6.39)$$

$$C_0^+ = 0$$

$$C_n^+ = \max(0, C_{n-1} + a(Y_n) - k^+) \quad (6.40)$$

and the corresponding downward CUSUM tests for increases in λ.

The pivotal $a(y)$ follows a chi-squared distribution with one degree of freedom, so once the reference value has been determined, the CUSUM for the shape parameter of the inverse Gaussian can be handled using the same methods as we developed for CUSUMs of gamma variables.

CUSUM schemes for the inverse gaussian mean appear in Nabar and Bilgi (1994), Olwell (1996), and Edgeman (1996). The scheme for the inverse Gaussian shape parameter λ appears in Olwell (1996).

6.2.8 *The Weibull distribution*

Another, perhaps more common, distributional model for lifetimes is the Weibull distribution, with density given by

$$f(y|\alpha,\beta) = \alpha\beta^{-\alpha}y^{\alpha-1}\exp\left(-(y/\beta)^{\alpha}\right), \tag{6.41}$$

where α is a shape parameter, and β is a scale parameter. The CUSUM scheme of most common interest is one for a scale change.

Examining Equation 6.41 with α fixed, we find:

$$\begin{aligned} a(y) &= y^{\alpha} \\ b(\beta) &= \beta^{-\alpha} \\ d(\beta) &= \alpha\ln(\beta). \end{aligned}$$

Following our now familiar principles, we have that the optimal CUSUM for an increase in β (change of scale) for a Weibull distributed random variable is given as follows.

$$k^{+} = \frac{\alpha\ln(\beta_1/\beta_0)}{\beta_1^{-\alpha} - \beta_0^{-\alpha}} \tag{6.42}$$

$$C_0^{+} = 0$$

$$C_n^{+} = \max(0, C_{n-1}^{+} + y^{\alpha} - k^{+}) \tag{6.43}$$

which signals when $C_n^{+} > h^{+}$. We select h^{+} to achieve a specified ARL in control.

Like the inverse Gaussian shape parameter, the Weibull scale can be handled the same way as a gamma variable. This happens because the distribution of Y^{α} is exponential with mean β^{α}, which means we can use the ARL routines we have developed for the gamma distribution.

Johnson (1966) developed the original, equivalent CUSUM scheme for the V-mask for the scale change of a Weibull distribution.

6.2.9 *Distributions outside the exponential family*

The optimal diagnostic for a step change in the parameter of any distribution is a CUSUM of the score statistic

$$Z_n = \ln\left(\frac{f(X_n|\theta_1)}{f(X_n|\theta_0)}\right).$$

This is true whether the distribution $f(x|\theta)$ is in the exponential family or not. What is special about the exponential family is that the score statistic is essentially the same whatever the values of θ_1 and θ_0: it is the sufficient statistic $a(X_n)$, with the in-control and out-of-control parameter values appearing only in the reference value.

For distributions outside the exponential family, the score statistic involves the two parameter values more intrinsically. Different choices of θ_1 change not just a constant reference value, but the actual score statistic. This means that non-exponential-family CUSUMs can have strange properties, for example, worse performance to a shift larger than that for which the CUSUM was designed.

As one example, the scaled Student's t distribution is sometimes used as a model for data following a symmetric distribution but with heavier tails than the normal. This distribution, with mean μ, can be written in the form

$$f(x|\mu, \alpha, \gamma) \propto \frac{1}{\{\alpha + (x - \mu)^2\}^\gamma}.$$

The optimal diagnostic to test whether the mean has shifted from an in-control level μ_0 to an out-of-control level μ_1 is a CUSUM of the scores

$$Z_n = \gamma \ln \left(\frac{\alpha + (X_n - \mu_0)^2}{\alpha + (X_n - \mu_1)^2} \right).$$

This score clearly involves the X_n and the in-control and out-of-control parameter values, but not in a simple separable way as was the case within the exponential family.

As CUSUMs outside the exponential family are quite specialized, we say no more about them here.

> This general method of finding CUSUM schemes for changes in a parameter may be used for any member of the exponential family of distributions. The CUSUM is of the sufficient statistic minus a reference value.

6.3 The Markov property of CUSUMs

In this section, we discuss just the upward CUSUM. For purposes of finding the ARL of a downward CUSUM, we just reverse the sign and turn it into an upward CUSUM also. In other words, the downward CUSUM

$$C_n^- = \min(0, C_{n-1}^- + X_n - k^-)$$

can be turned into the upward CUSUM

$$D_n^+ = \max(0, D_{n-1}^+ + (-X_n) - (-k^-))$$

(where $D_n^+ = -C_n^-$) of a random variable that is the negative of X_n and with an allowance whose sign is reversed.

As we only consider upward CUSUMs, to simplify notation we suppress the "+" superscript and just refer to the upward CUSUM as S_n.

The CUSUM has the Markov property: given the nth value of the CUSUM, the previous values have no effect on the $n+1$th value of the CUSUM. More precisely:

$$P(S_n = s_n | S_0 = s_0, S_1 = s_1, \ldots, S_{n-1} = s_{n-1}) = P(S_n = s_n | S_{n-1} = s_{n-1}) \tag{6.44}$$

This follows immediately from the independence of the sampled observations Y and the recursive definition of S_n.

The CUSUM for continuous random variables has a continuous state space; S_n takes values from an interval of real numbers. The CUSUM for discrete random variables has a discrete state space. In both cases, the CUSUMs are discrete time Markov processes: the process is indexed by the sample number, which is a member of the nonnegative integers.

The study of continuous space discrete time Markov processes involves significant probability theory. Doob (1953) is the classic reference.

For our purposes, we can approximate the continuous state space of the CUSUM of a continuous random variable by a partition of the interval $(0, h)$ into M equal subintervals, making each of these subintervals a state in the Markov process, along with one state for $S_n = 0$ and another for $S_n \geq h$. These $M + 2$ states allow us to consider the CUSUM as a Markov chain.

This simplification allows us to apply the many results available for Markov chains to problems involving the distribution of S_n. These discrete approximations increase in accuracy as M grows, but can be accurate in many cases with quite manageable values for M.

Brook and Evans (1972) first introduced the Markov chain model for CUSUMs.

Using a mesh of width $\Delta = h/M$, we discretize the range of possible S values into the states:

$$\begin{array}{rcl} \text{State 0} & S = 0 & \\ \text{State i} & S \in ((i-1)\Delta, i\Delta] & i = 1, 2, \ldots, M \\ \text{State M + 1} & S > h & \end{array}$$

We define the transition probability matrix T with elements

$$t_{i,j} = P(S_n \in \text{State j} | S_{n-1} \in \text{State i})$$

for $i, j = 0, 1, \ldots, M + 1$. From State $M + 1$ we transition to State 0 with probability 1. (In other words, when the CUSUM exceeds its decision interval it will be restarted.)

We discuss methods for determining T in the following.

It is well known that the mean return time between visits to a recurrent state for an irreducible Markov chain is the reciprocal of the stationary probability for that state (Hoel et al., 1972). The ARL for the process as we have modeled it is the average time between visits to state $M+1$, the state corresponding to $S_n > h$, with one modification. The transition from State $M+1$ to State 0 occurs instantaneously, so we must adjust our ARL. Accordingly, the ARL is the reciprocal of the stationary probability that the process is in state $M+1$ minus one, or

$$ARL = \frac{1}{\pi_{M+1}} - 1. \tag{6.45}$$

The eigenvector of the matrix T associated with the eigenvalue 1 gives the stationary distribution for the Markov chain (Hoel et al., 1972). This eigenvector is easily extracted through numerical methods, and from it we can calculate the ARL.

6.4 Getting the ARL

We have derived the optimal CUSUM scheme, finding the optimal statistic $a(Y)$ and reference value k. We stated that the value of the parameter h, the decision interval, depended on the desired in-control ARL. In this section we address three methods for determining the ARL of a CUSUM scheme for a given value of h. These methods form the basis for algorithms to find the value of h for a given ARL, the inverse problem.

The three methods are integral equations, discrete approximations to the integral equations based on Markov chains, and simulation schemes that take advantage of variance reduction techniques. The discrete approximation based on Markov chains forms the basis for the computer algorithms on the Web site.

6.4.1 The renewal equations

Page (1954) introduced integral equations for determining the ARL of a CUSUM in his original article. Van Dobben de Bruyn (1968) also discussed them.

We use the following notation. $L(z)$ is the expected run length or ARL for the CUSUM that begins at $S_0 = z$. Usually, we are interested in $L(0)$, but the fast initial response (FIR) scheme of Lucas and Crosier (1982b) motivates the study of $L(z)$ for nonzero head start values. We discussed these head starts in Chapter 3.

$F(x)$ is the distribution function of the CUSUM statistic $X = a(Y)$. Assuming F has a density, we write it as $f(x)$.

The ARL from any point is equal to the sum of four components, the last three corresponding to whether the next observation puts the CUSUM equal to zero, somewhere in the interval $(0, h)$, or greater than h. These three choices exhaust the possibilities. We write the ARL for the current value of z in terms of the next value of z.

The first component is one, because we always draw at least another observation for $z \in (0, h)$.

The second value is the probability that the next observation returns the CUSUM to the zero, multiplied by the ARL from zero. This probability is $F(k-z)$, since $S_n + a(y) - k < 0$ implies that $a(y) < k - S_n = k - z$.

The third is the integral of the ARL for the next value of the CUSUM if it is between 0 and h times the probability that this next value occurs. This integral is given by

$$\int_0^h L(x)dF(x+k-z). \tag{6.46}$$

If the density exists, we can write Equation 6.46 as

$$\int_0^h L(x)f(x+k-z)dx. \tag{6.47}$$

The last value is the ARL from the next value of the CUSUM, if that next step of the CUSUM results in $S_n > h$. That ARL is of course zero, so this term drops out.

Putting the pieces together, we have the following integral equation for continuous variables:

$$L(z) = 1 + L(0)F(k-z) + \int_0^h L(x)f(x+k-z)dx. \tag{6.48}$$

We are often most interested in the values for $L(0)$. However, from the form of Equation 6.48, it is clear that to find $L(0)$ we must know $L(z)$ on the interval $[0, h)$. We get the rest of the values of $L(z)$ "for free." As mentioned before, these values are useful for designing and analyzing FIR CUSUMs.

The solution of the integral equations can be found symbolically for certain distribution functions, such as the exponential distribution (Van Dobben de Bruyn, 1968). More frequently, the equation must be solved using either numerical approximations or Markov chain approximations. We discuss the Markov chain approximations next.

6.4.2 The Markov chain approach

To implement our Markov chain approaches, we need to find the transition probabilities for our transition matrices. In other words, we need a general way to find

$$P(a < S_n < b | c < S_{n-1} < d). \tag{6.49}$$

The approach of Brook and Evans (1972) was to place S_n and S_{n-1} at the center of each interval. We prefer a more accurate approximation of the transition probabilities, based on Hawkins (1992b).

Let $\mu(x)$ be the distribution function of S_{n-1} conditional on $c < S_{n-1} < d$. As before, we write the distribution function of the sufficient statistic $X = a(Y)$ as $F(x)$.

Then we have

$$P(a < S_n < b \mid c < S_{n-1} < d) = \int_c^d \{F(b - s + k) - F(a - s + k)\} d\mu(s). \tag{6.50}$$

A priori, we do not know the measure μ, but we can substitute some reasonable measure. Hawkins (1992b) used the uniform distribution after exploring some other choices without finding a better one.

We can further approximate the integral in Equation 6.50 using Simpson's rule for the interval $[c, d]$ with midpoint m. This gives the approximation

$$\begin{aligned} P(a < S_n < b|c < S_{n-1} < d) = \\ [\{F(b - c + k) + 4F(b - m + k) + F(b - d + k)\} \\ -\{F(a - c + k) + 4F(a - f + k) + F(a - d + k)\}]/6. \end{aligned} \tag{6.51}$$

Equation 6.51 can be applied repeatedly to give accurate values for the transition matrices discussed earlier in the chapter.

Given T, how do we find the ARL? We use the matrix R, which disregards the transitions to and from the last state. For this Markov chain, we let z be one of the $M + 1$ first states. Then we have

$$L(z) = 1 + \sum_{i=0}^{M} L(i) R_{i,z}. \tag{6.52}$$

(We omit the last state because the ARL from State $M + 1$ is known to be zero.)

Equation 6.52 is exactly the discrete analog of the integral equation given by Equation 6.48. We can write Equation 6.52 in matrix form. Let λ be the vector of length $M + 1$ of ARL values for CUSUMs starting in the corresponding states $0, 1, \ldots, M$, and let $\mathbf{1}$ be a vector of length $M + 1$ all of whose elements are 1. Then

$$(I - T)\lambda = \mathbf{1} \tag{6.53}$$

This matrix representation is one approach to finding ARLs. Solving Equation 6.53 involves finding a solution to $M + 2$ linear equations.

An alternate approach to finding the ARL from the matrix T involves finding the stationary distribution for the Markov chain. Since T is irreducible and finite, a unique stationary distribution exists (Hoel et al.,

1972). A crude but simple method involves finding successive squares of T until the stationary distribution can be read from the result. Since T has one eigenvalue equal to one, and all others have magnitude less than 1, $\lim_{n\to\infty} T^n = [\pi, \pi, \ldots, \pi]^T$, where π is the stationary distribution. Taking successive squares of $T : T^{2^{2^{\cdots}}}$ quickly and easily approximates this limit matrix. We can then read the stationary distribution right from the limit matrix.

The algorithms presented later in this work use the first matrix formulation, based on the matrix T and Equation 6.53, to find the ARL.

6.4.3 Simulation using variance reduction techniques

There is another approach to determining ARLs. One can simulate the process producing the CUSUM, record run lengths, and then average to obtain the ARL. The collection of run lengths can also be used to derive an empirical distribution of the run lengths. For processes with long ARLs, this crude simulation can be computationally intensive and inefficient.

Jun and Choi (1993) developed two variance reduction schemes for improving estimates of the ARL using simulation. The first uses the total hazard as a control variate, as proposed by Ross (1990). The second approach uses ratio estimators based on the length of cycles, where a cycle is completed if the CUSUM resets at the origin or if it signals. Since there are usually many cycles in a single simulation run, this method results in a variance reduction.

Jun and Choi provide examples showing that improvement over crude simulation of one to four orders of magnitude in the precision of the estimates for ARL can be obtained with their methods.

We find these simulation methods most useful as checks on our implementation of the Markov schemes.

6.5 Summary

We have shown how CUSUM schemes follow from SPRT tests, and how the SPRT produces the reference value, k, for the CUSUM. We showed the CUSUM scheme for an arbitrary member of the exponential family of distributions, and explicitly found the schemes for several common members. We discussed the optimality properties of the CUSUM. We closed with a discussion of methods to determine the *ARL* of a CUSUM scheme.

The theory in this chapter underpins the results of the first five chapters of the text.

6.6 Further reading

Barnard (1959) developed control charts for a process whose mean undergoes step changes according to a Poisson process. This puts some additional structure on the out-of-control situation whose occurrence is not usually modeled explicitly in CUSUM work.

The assumption that θ is equal to either the single in-control value θ_0 or a single known out-of-control value θ_1 is usually an oversimplification. An alternative is to form the maximum likelihood estimator $\hat{\theta}_1$ of observations $k \ldots n$ and use it in the likelihood ratio test of the null hypothesis that $\theta_1 = \theta_0$. This forms the basis of the "generalized likelihood ratio," or GLR schemes which are discussed in detail in Basseville and Nikiforov (1993). Lorden (1971) showed that both the in-control and out-of-control ARL of this scheme are of the same order as those of the simple fixed-parameter case. Siegmund (1995) derived close approximations to the average run length of the generalized likelihood ratio statistic.

Lai (1995) discusses many solved and open problems for series that depart from the simple independence model. One generally useful method is that of the "window-limited GLR scheme" in which we form the generalized likelihood ratio test statistic for a change, but instead of evaluating it over the entire data sequence, limit the calculations to some finite window backward from the current reading. This device prevents the calculation from getting out of hand, but turns out to have only a modest impact on chart performance.

Goel and Wu (1971) and Fellner (1990) used the renewal theory approach to calculate the ARL of the normal mean CUSUM, providing nomograms and a general-purpose program, respectively.

There is often use for a computationally simple approximation to the ARL. Reynolds (1975) used a Brownian motion approach to get approximations for a CUSUM of normal data. Siegmund (1985, equation 2.57) gave a particularly simple approximation for the ARL of a CUSUM of normal data. Consider the upward CUSUM of standard normal $N(0,1)$ data, with reference value k and decision interval h. Siegmund's approximation for the ARL is

$$ARL \approx \frac{e^{-2k(h+1.166)} + 2k(h+1.166) - 1}{2k^2}.$$

This approximation is accurate to within a few percent for small to moderate values of k (such as $k \leq 1$) but can be quite inaccurate for larger values of k.

Siegmund (1985, Chapter 10) also gives a general method of finding run length approximations for any member of the exponential family, but this more general approach does not appear to have been exploited extensively.

We have concentrated on the run lengths of a one-sided CUSUM. The ARL of a two-sided CUSUM can be found from those of various one-sided

CUSUMs provided the upward and downward CUSUMs do not interact. A Markov chain joint model for the upward and downward CUSUMs is given by Woodall (1984).

7
Calibration and short runs

It is a capital mistake to theorize before one has data.
A. Conan Doyle

Introduction

Up to now, we have not said much about the estimation of the in-control parameters, except to point out that large samples are necessary to estimate them adequately. To get some feeling for this, we try to quantify the impact of uncertainty in the parameter estimates in the case of a normal CUSUM. Suppose that the process stream is $N(\mu, \sigma^2)$. To calibrate the CUSUM, we take a sample of size m and compute its mean $\overline{X}$ and standard deviation s, and substitute $\overline{X}$ for μ and s for σ. This means that our figure for the true mean is in error by the amount $\mu - \overline{X}$, and the figure for the true standard deviation is in error by the ratio σ/s.

We do the calculations assuming the standard normal $\mu = 0, \sigma = 1$. Some thought (or a little algebra) shows that the actual values of μ and σ are irrelevant — all that matters is m, the size of the calibration sample.

Suppose a sample of size $m = 50$ is used for the calibration. Then the standard error of $\overline{X}$ is

$$se_{\overline{X}} = \frac{1}{\sqrt{m}} = 0.141,$$

		$\overline{X}$		
		-0.14	0	0.14
	0.9	156	411	1223
s	1.0	319	999	3541
	1.1	720	2883	11439

TABLE 7.1. In-control ARL of CUSUMs with estimated parameters

and the standard error of s is approximately

$$se_s = \frac{1}{\sqrt{2m}} = 0.1.$$

There is a 2/3 chance that $\overline{X}$ will be within 1 standard error of μ, and a 1/6 chance of its being more than 1 standard error above μ, and a 1/6 chance of being more than 1 standard error below μ. So an error of more than a standard error is quite likely. The sample is large enough that the standard deviation s is roughly normally distributed, so it can be interpreted very similarly.

Let's turn these figures into 3 scenarios for $\overline{X}$: that $\overline{X}$ equals μ; that $\overline{X}$ is 1 standard error above μ; and that $\overline{X}$ is 1 standard error below μ. We also set up the same three scenarios for s. Combining all possibilities gives 9 scenarios. We design a one-sided CUSUM for upward shift in mean to have an in-control ARL of 1,000, setting $k = 0.5, h = 5.07$, and see what the actual ARL is under each of these scenarios.

If $\overline{X}$ is 1 standard error above μ — that is, at 0.14, and s is 1 standard error above σ — that is, at 1.1, then the in-control data look as though they have a mean shift of $\Delta = -0.14/1.1 =$ -0.128 standard deviations, and the true standard deviation itself appear to be a fraction $\lambda = 1/1.1 = 0.909$ of what it should be. We can then calculate the ARL of the CUSUM using the results of Chapter 3. Doing this calculation for each of the 9 possibilities gives the ARLs in Table 7.1

The figures in Table 7.1 are quite shocking. They show nearly two orders of magnitude difference between the ARL of 11,439 in the bottom right corner and that of 156 in the top left corner.

If the CUSUMs are two-sided the effect is less dramatic, since the change in ARL on the downward CUSUM would offset that on the upward CUSUM, but even looking just at the 3 scenarios with $\overline{X} = \mu$, there is a factor of 7 from the high ARL to the low.

Clearly, the unavoidable random variability in the estimates of μ and σ based on a calibration sample of size 50 has an effect on the CUSUM's ARL that is so large as to be a cause for much concern.

We should mention that this sensitivity to errors in the parameter estimates does depend on the particular CUSUM design. A CUSUM with small k (and so designed to detect small shifts) is much more sensitive to random errors in the parameter estimates than a CUSUM with a large k, which by implication is looking for larger shifts. At the extreme, the Shewhart chart, which is the same as a CUSUM with $k = 3, h = 0$ is not very sensitive to these random errors, and so can be calibrated with quite modestly sized samples. The choice $k = 0.5$ used here is a "middle of the road" choice, and really small values like $k = 0.2$ would be affected much more severely by the random error in parameter estimates.

The obvious remedy to this sensitivity to the parameter estimates is the one put forward in Chapter 1: to use a larger calibration sample. Doubling the size of the calibration sample from 50 to 100 would stabilize the ARL considerably, but even in this case the lowest ARL among the 9 scenarios is 300 and the highest 4,200. This range is certainly better than what we got from a calibration sample of size 50, but it is in no sense good.

We could demand even larger calibration samples. Not only may this asking for long data sets just to calibrate the SPC be undesirable, it may be impossible. In some processes, long runs are the exception rather than the norm, and we have to find some way to live within the confines of more moderate calibration samples.

We could, in principle, make allowance for the fact that parameters are estimated using standard statistical methods. Suppose we calibrate a CUSUM using the sample mean $\overline{X}$ and standard deviation s of a sample of size m. Then the distribution of the standardized quantities used to make the CUSUM,

$$U_n = (X_n - \overline{X})/s,$$

rather than following the $N(0, 1)$ distribution of exactly known parameters, would follow a scaled Student's t distribution. We could incorporate this exact distribution, together with the fact that the successive U_n are interdependent because of their use of a common $\overline{X}$ and s, into an exact CUSUM ARL calculation, but this would not be an easy task.

> Unavoidable variability when estimating the parameters of an in-control process can greatly affect the ARL of the standard CUSUM.

7.1 The self-starting approach

There is another possible approach, and one that removes the estimation issue from the problem completely. This is based on the idea of using the regular process measurements themselves for both the purposes of calibrating the CUSUM and maintaining control. This leads to "self-starting" CUSUMs in which each successive observation is standardized using the mean and standard deviation, not of a special calibration sample, but of all observations accumulated to date. This means that it is not necessary to assemble a large calibration data set before the control begins (although it is generally advisable to collect a few preliminary observations). Then as the process continues to run and produce additional observations, the estimates of the mean and standard deviation get closer and closer to the true values.

By using appropriate statistical distribution theory that recognizes that we are using estimates rather than the true parameter values for μ and σ, we can account for the random variability in the parameter estimates and get the intended ARLs.

Using ongoing routine process data does expose self-starting CUSUMs to the danger of contaminating these running means and standard deviations by including data from out-of-control states. Some precautions are needed to protect against this happening and we sketch these as the discussion advances.

7.2 The self-starting CUSUM for a normal mean

We start with the earliest and most widely useful self-starting CUSUM, that for control of a normal mean. As usual, we suppose that the process readings have an in-control $N(\mu, \sigma^2)$ distribution, but now these true values are not assumed known.

Write $\overline{X}_n$ for the mean of the first n process readings, and let $W_n = \sum_{j=1}^{n}(X_j - \overline{X}_n)^2$ be the sum of squared deviations of the first n readings from their mean. The (sample) variance of the first n readings is then given by $s_n^2 = W_n/(n-1)$.

Standardizing each reading using the running mean and standard deviation of the preceding observations gives:

$$T_n = \frac{X_n - \overline{X}_{n-1}}{s_{n-1}}$$

which is defined for n values of 3 or greater (two readings being required to get an initial mean and standard deviation). Under the normal distribution assumption, T_n follows a scaled Student's t distribution:

$$\sqrt{\frac{n-1}{n}}T_n \sim t_{n-2}.$$

The exact cumulative distribution function of T_n is then given by

$$Pr[T_n < t] = F_{n-2}\left(t\sqrt{\frac{n-1}{n}}\right).$$

where $F_{n-2}(.)$ stands for the cumulative distribution function of the Student's t distribution with $n-2$ degrees of freedom. Note that we use F here for the cumulative distribution function, *NOT* the Fisher F distribution.

As $n \to \infty$, the t distribution gets closer and closer to the standard normal distribution. When textbooks discuss the t distribution and its near normality for large degrees of freedom, they usually say this happens because for large n, $\overline{X}$ is close to μ and s is close to σ, but this is a major oversimplification.

The reality is that in moderate to large samples, the random errors in s are sufficiently close to a symmetric distribution that during repeated sampling the overestimates of σ almost exactly cancel out the underestimates. When doing two-sample tests with large degrees of freedom, it is standard practice to use the normal distribution as an approximation to the t. This works for single samples, but it will not do for the repeated sampling involved in CUSUM testing. This is because it ignores the correlation between the T_n created by use of a common estimate for μ and for σ. Getting the correct ARL of a CUSUM using sample estimates of the process parameters needs a more accurate distributional analysis.

Knowing the exact distribution of T_n allows us to transform it to a quantity that is exactly standard normal for all sample sizes. The basis for this is the general result that converting any continuous random variable to its tail area and then converting that tail area to a normal ordinate defines a random variable that is exactly $N(0,1)$. If we write Φ^{-1} for the inverse normal function, that is, the function taking a normal area and producing the ordinate that has that area to the left of it, then the transformation

$$U_n = \Phi^{-1}[F_{n-2}(a_n T_n)]; \qquad a_n = \sqrt{\frac{n-1}{n}}$$

will transform the "studentized" CUSUM quantity T_n into a random variable U_n that has an exact $N(0,1)$ distribution for all $n > 2$.

A final key piece of the scheme is a lemma due to Basu (Dawid, 1979), which shows that the U_n are statistically independent. Thus, even without a huge calibration sample to damp down the sampling variability in $\overline{X}$ and s, by transforming the T_n to their U_n counterparts we still get a sequence of independent $N(0,1)$ values to CUSUM.

The self-starting CUSUM scheme then is:

n	X_n	$\overline{X}_n$	W_n	s_n	T_n	a_nT_n	$F_{n-2}(a_nT_n)$	U_n
1	57	57.00	0	—	—	—	—	—
2	65	61.00	32	5.66	—	—	—	—
3	60	60.67	33	4.04	-0.18	-0.14	0.4557	-0.11
4	53	58.75	77	5.06	-1.90	-1.64	0.1213	-1.17
5	70	61.00	178	6.67	2.22	1.99	0.9297	1.47
6	50	59.17	279	7.47	-1.65	-1.51	0.1028	-1.27
7	60	59.29	279	6.82	0.11	0.10	0.5379	0.10
8	75	61.25	495	8.41	2.30	2.15	0.9625	1.78
9	58	60.89	505	7.94	-0.39	-0.36	0.3647	-0.35
10	75	62.30	684	8.72	1.78	1.69	0.9353	1.52

TABLE 7.2. Calculations for self-starting CUSUM.

- For each n, compute $\overline{X}_n$, W_n (and from that s_n).
- For $n > 2$, compute T_n and its transform U_n.
- Form a CUSUM of the U_n. This CUSUM, being a CUSUM of exactly $N(0,1)$ independent quantities, will have the exact in-control ARL properties of normal data with known true mean and variance.

It may be helpful to see an example of the calculations, and Table 7.2 shows the steps in going from X_n to U_n. The successive columns are the running mean, the running sum of squared deviations from the mean, the running standard deviation, the studentized deviation of each observation from the preceding sequence, its Student t equivalent, P value, and finally equivalent normal score.

Once the U_n are generated, they can be handled exactly as outlined in Chapters 2 and 3. For example, they may be accumulated in V-mask form CUSUMs or in decision interval CUSUMs. They may also be used to make scale CUSUMs.

The calculations of the running mean and variance are considerably simplified by the fact that they can be written using the updates:

$$\overline{X}_n = \overline{X}_{n-1} + (X_n - \overline{X}_{n-1})/n$$

$$W_n = W_{n-1} + (n-1)(X_n - \overline{X}_{n-1})^2/n$$

so the updates can be made very rapidly. If the computations were done by hand, getting the t tail areas $F_{n-2}(a_nT_n)$ would be a difficult chore, but this is made trivial by doing the calculations on a computer using standard procedures for t tail areas and the inverse normal.

We provide an Excel spreadsheet *SSNORMAL.xls* from the Web site which offers an easy convenient way of generating these self-starting CUSUMs for both location and scale for the normal distribution.

We have concentrated on using the U_n to construct a CUSUM chart. They also provide us with an immediate Shewhart chart for checking the compatibility of each observation with all those preceding it.

This Shewhart chart is valuable for locating special causes that do not persist, and is an important complement to the self-starting CUSUM. We may plot individual U_n values with a center line of 0 and control limits at 3 and -3 (or other limits to achieve a particular in-control ARL). Or, if there is some natural rational grouping of the data (for example, if we are taking a number of observations in the same shift), we could average the U_n within rational groups and plot these rational group means on a Shewhart chart.

> The self-starting CUSUM scheme provides a self-starting Shewhart chart for free.

7.2.1 Special features of self-starting charts

There are some major differences between self-starting CUSUMs and their known-parameter counterparts. The most important of these revolve around the out-of-control situation. Think about a known-parameter decision interval CUSUM for the mean of normal data following an upward shift in mean. Provided the shift is bigger than the reference value, the DI CUSUM tends to move upward indefinitely, centered on a straight line of slope $\Delta - k$. This does not happen with a self-starting CUSUM. If the CUSUM is allowed to continue without interruption, immediately after the shift in mean it initially move upward, just like the known-parameter CUSUM. But then as the shifted values are fed into the running mean and variance, they move the running mean up toward the new mean, and they also inflate the variance. The combined effect of these two properties is that if the DI CUSUM is left to run unhindered after initially rising, it turns back down below the decision interval.

To get some feeling for these effects, imagine that the process runs in control at $N(\mu_0, \sigma^2)$ for a total of m_0 readings and then shifts to a $N(\mu_1, \sigma^2)$ distribution. After m_1 readings in the new regime, the expected value of the running mean and sum of squared deviations are (using the analysis of variance identity)

$$\mu_0 + \frac{m_1}{m_0 + m_1}(\mu_1 - \mu_0)$$

$$(m_0 + m_1 - 1)\sigma^2 + \frac{m_0 m_1}{m_0 + m_1}(\mu_1 - \mu_0)^2,$$

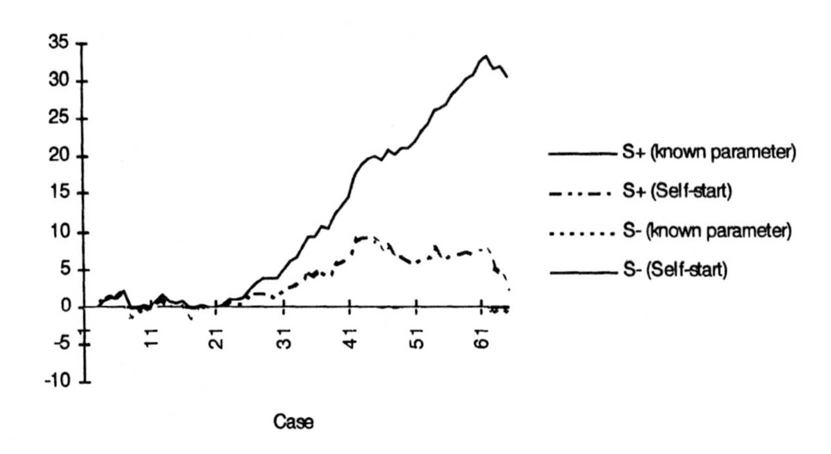

FIGURE 7.1. Self-starting CUSUMs and the out-of-control state.

respectively. Note that a large shift in mean can ramp the running variance up sharply, while the running mean will move up tangentially toward the new level. Following the shift in mean, the deviations $X_n - \overline{X}_{n-1}$ will have expectation

$$\frac{m_0}{n-1}(\mu_1 - \mu_0)$$

which move toward zero, making the shift less and less easily detected against the background of the running mean. This tendency for the CUSUM to adapt to the new mean will be largest when m_0 is small: that is, when there is only a short period of in-control behavior for the CUSUM to "learn" the in-control parameters before the process goes out of control. The longer the in-control period m_0 before the shift, the more slowly will the self-starting CUSUM lose its ability to detect the shift.

Figure 7.1 illustrates this effect. To produce it, we generated a random sample in which the first 20 cases were $N(50, 25)$ and the remainder were $N(55, 25)$: in other words, had a one standard deviation shift after the first 20 observations. We put these data through two decision interval CUSUMs — a known-parameter CUSUM, and a self-starting CUSUM — both using $k = 0.5$. These CUSUMs show the effect quite clearly. Both CUSUMs move up strongly following the shift at observation 20, but the known-parameter CUSUM carries on moving upward throughout the series, and the self-starting CUSUM flattens and seems to be moving back down.

Both CUSUMs detect the change quickly: the known-parameter CUSUM crosses the boundary $h = 5.07$ at observation 32, or 12 readings after the actual shift, and the self-starting CUSUM does so at observation 39, or 19 readings after the shift.

The danger in the CUSUM flattening out and even coming back below the decision interval is that a user unfamiliar with self-starting CUSUMs might assume this meant the problem had gone away by itself, whereas it could mean no more than that the running mean had almost caught up with the new shifted mean. A more alert user will act on the signal immediately and look for the special cause giving rise to it.

Signals from self-starting charts must be acted upon immediately

When a self-starting CUSUM gives a signal, it is essential to remove the "contaminated" data from the running mean and standard deviation as part of the corrective action of diagnosing the special cause. This can be done by rolling the history back to the estimated last in-control value (the last reading at which the DI CUSUM was on the axis) and restarting the self-starting CUSUM from this point. In the simulated data, this would lead to rolling the history back to reading 22, at which stage the running mean and standard deviation were 50.91 and 4.69, respectively. At the time of the signal, the running mean was up to 53.2, most of the way toward the new mean being contaminated with nearly as many shifted as in-control observations.

In the worst case, the process mean might shift very soon after the self-starting CUSUM was started, and then the running mean would quickly adapt to the new level, quite possibly before there was enough time to signal the shift. It is for this reason that it may be prudent to start the CUSUM out with a few of the sort of carefully-watched process readings commonly used to calibrate control charts. We do not need to have hundreds or even dozens of these readings;the example of Figure 7.1 illustrated that the number of readings needed to protect against an immediate large shift is around a dozen or two.

The other thing that will derail the self-starting CUSUM is an outlier in the data. This may not trigger a signal on the CUSUM (particularly if the CUSUM is tuned for a small shift in mean and therefore has a large h), but can wreak havoc on the running standard deviation. Once the standard deviation is inflated, the potential for identifying mean shifts or further outliers (either of which could compound the problem) is further weakened. Thus it is vital to protect the running mean and standard deviation from the introduction of outliers. There are many ways of doing this. One that generally works well is "metric winsorization." Look at the update formulas for the running mean and sum of squared deviations:

$$\overline{X}_n = \overline{X}_{n-1} + d_n/n$$

$$W_n = W_{n-1} + (n-1)d_n^2/n,$$

where d_n is the deviation of X_n from the running mean

$$d_n = X_n - \overline{X}_{n-1}.$$

An outlying X_n could make this deviation arbitrarily large — positive or negative — and thereby distort $\overline{X}_n$ and W_n arbitrarily severely.

Metric winsorization reduces the potential damage from outliers by limiting the possible size of d_n, redefining it as

$$d_n = \begin{cases} -C_n, & X_n - \overline{X}_{n-1} < -C_n \\ X_n - \overline{X}_{n-1}, & -C_n < X_n - \overline{X}_{n-1} < C_n \\ C_n, & X_n - \overline{X}_{n-1} > C_n. \end{cases} \tag{7.1}$$

The constant C_n is a winsorizing constant, typically set to a value in the range 1 to 3 standard deviations. We return to the topic of winsorization later. It is valuable in known-parameter CUSUMs also, but for a different purpose. There it is used to help distinguish situations where the process parameter shifts and stays shifted from those where it deviates sharply but then returns on its own to the in-control state. This second use of winsorization is useful in self-starting CUSUMs also, where we can define the standardized deviation as $T_n = d_n/s_{n-1}$, but it is the first use of protecting the parameter estimates from contamination that makes winsorization particularly valuable in self-starting analysis.

Another difference between self-starting and known-parameter CUSUMs is that interpretation following a signal is also not as easy with self-starters as in the known-parameter case. In the known-parameter CUSUM, the out-of-control CUSUM centers on a straight line whose slope estimates the shift in mean. A self-starting DI CUSUM starts out roughly linear following a parameter shift, but after that curves over and moves back toward the axis. Thus estimation of the new parameters requires arithmetic rather than geometry when using self-starting CUSUMs: you calculate the before-and-after means and standard deviations using records of the running means and standard deviations. You cannot calculate them using the slope of the CUSUM.

To summarize, a self-starting CUSUM does not require a long in-control calibration sequence to estimate the process parameters. If one is available anyway, then the self-starting CUSUM will work better than the traditional method of plugging in the calibration mean and standard deviation for the known parameter values.

As a detection device, the self-starting CUSUM is a no lose proposition.

7.3 Self-starting CUSUMs for gamma data

The key theoretical features that allowed the self-starting CUSUM to work so well for normal data were:

- the normal distribution is in the exponential family, which makes it easy to update the maximum likelihood estimates of its parameters case by case as new observations become available;
- the normal distribution has a standardized "pivotal quantity" $(X_n - \mu)/\sigma$ whose distribution is the same for all μ and σ values;
- the estimates $\overline{X}_n$ and s are complete and sufficient, which is enough to show, using the Basu lemma, that the standardized quantities $(X_n - \overline{X}_{n-1})/s_{n-1}$ are mutually independent.

The same sort of theoretical properties hold for other distributions in the exponential family, and allow us to define self-starting approaches for CUSUMs of these distributions as well. For example, the gamma (and so also the chi-squared) distributions are an important family that also fits this framework.

7.3.1 *Background*

The gamma distribution has density function (given earlier by Equation 6.6)

$$f(x) = \frac{x^{\alpha-1}e^{-x/\beta}}{\Gamma(\alpha)\beta^{\alpha}}.$$

We use the shorthand $\Gamma(\alpha, \beta)$ for this distribution. In the problems of common statistical interest, the shape parameter α is known. For example, when we CUSUM sample variances, α is half the number of degrees of freedom, and as the size of the rational groups is known, so is α.

In the known-parameter case, controlling the gamma distribution for (upward or downward) shifts in its scale parameter β from an in-control level β_0 to an out-of-control level β_1 is done using the optimal CUSUM of Equations 6.13 and 6.14:

$$C_0 = 0$$

$$C_n = \max\left(0, S_{n-1} + X_n\left(\beta_0^{-1} - \beta_1^{-1}\right) + \alpha\ln(\beta_0/\beta_1)\right).$$

X_n has mean $\alpha\beta$, so an increase in β corresponds to an increase in the mean of X_n.

The gamma distribution has a standardized form. This uses the pivotal

$$U_n = X_n/\beta$$

which follows the $\Gamma(\alpha, 1)$ distribution.

The maximum likelihood estimate of β, with α known, based on a sample is

$$\hat{\beta} = \frac{\overline{X}}{\alpha},$$

the arithmetic mean of the observations divided by α. The estimate is sufficient and complete; this is all that is needed to set up a self-starting CUSUM scheme for gamma-distributed quantities.

7.3.2 The scheme

Write $\overline{X}_n$ for the arithmetic mean of the first n of the process readings. Using the best running estimate for β in the pivotal U_n gives the test quantity

$$F_n = X_n / \overline{X}_{n-1}$$

for all n values 2 or greater. This follows an exact Fisher's F distribution with 2α numerator and $2\alpha(n-1)$ denominator degrees of freedom.

Since the exact distribution of F_n is known, we can transform it to any convenient standard distribution, much as we did with the CUSUM for normal data.

One particularly attractive possibility is to transform it to a $\Gamma(\alpha, 1)$ distribution, the distribution of the CUSUMmand U_n that would give the optimal diagnostic for shifts in β if β_0 were known. We get a transform with this distribution by turning the observed F_n into a tail area, and then converting this tail area to the corresponding $\Gamma(\alpha, 1)$ value.

In symbols, write $F_{2\alpha,2\alpha(n-1)}(f)$ for the cumulative probability up to f of an F distribution with 2α and $2\alpha(n-1)$ degrees of freedom. Write $G_{\alpha}^{-1}(a)$ for the function that takes an area a and returns the value that has area a to the left of it under a $\Gamma(\alpha, 1)$ distribution. Then the transformation

$$U_n = G^{-1}\left[F_{2\alpha,2\alpha(n-1)}(F_n)\right] \tag{7.2}$$

will produce a sequence of U_n values that are exactly $\Gamma(\alpha, 1)$ distributed. Furthermore, Basu's lemma shows that they will be mutually independent. This means that these values can be used in CUSUMs in place of the ideal pivotals X_n/β_0 that would be used if the in-control value of β_0 were known exactly.

As the series length increases, the values of $\overline{X}_n/\alpha$ will converge to β, and so the exact transforms U_n will be closer and closer to the starting transforms F_n based on the sample estimate of β. Thus this self-starting approach to gamma-distributed data can be used without any long calibration series, yet once a long in-control history has built up, will have the same performance as the idealized known-parameter CUSUM.

n	X_n	$\overline{X}_n$	F_n	Area	U_n
1	2.11	2.11	—	—	—
2	4.40	3.26	2.08	.7528	5.42
3	1.50	2.67	.46	.2359	1.85
4	2.55	2.64	.95	.5330	3.57
5	2.85	2.68	1.08	.6015	4.06
6	.57	2.33	.21	.0719	.88
7	5.30	2.75	2.28	.9093	8.02
8	14.66	4.24	5.32	.9974	16.37
9	7.42	4.60	1.75	.8363	6.52
10	9.77	5.11	2.13	.9023	7.84

TABLE 7.3. Calculation of a self-starting gamma CUSUM for β, with α known.

7.3.3 *Example*

To illustrate the calculations, consider the data in Table 7.3, which simulate the variances of normal samples of size 5 which therefore have 4 degrees of freedom, giving $\alpha = 2$.

The table shows the successive steps in standardizing each X_n to the average of its predecessors, turning this ratio into an F distribution area, and then converting this area back to an equivalent χ^2 variate with 4 degrees of freedom.

This self-starting CUSUM for sample variances has the same broad properties as the one for normal location. It can detect shifts that occur quite soon after start up provided the shift is moderately large. Naturally, it works best if there is a long period of in-control operation before the shift, but even this desirable longer learning period is much shorter than the calibration period you would need to be comfortable with substitution of the sample estimate of β_0 and using the known-parameter CUSUM.

Figure 7.2 illustrates the operation of the self-starting CUSUM. For comparison, it also shows the known-parameter CUSUM of the same data. The data (the first 10 cases of which are the numbers in Table 7.3) were randomly sampled from a χ^2_4 distribution, but the last 50 cases' values were scaled up by a factor of 1.5 to simulate a variance increase. The reference value was chosen as $k = 4$; a decision interval of $h = 23.92$ would give an in-control ARL of 1,000.

As the figure shows, the self-starting CUSUM and the known-parameter CUSUM were almost equal right through the in-control period, and also for a while after the shift in the underlying variance. Well over to the right of the series, however, the known-parameter CUSUM continues to move up linearly, whereas the self-starting increasingly lags behind as the shifted variances feed into $\overline{X}_n$ dragging it upward and the summands F_n downward. In fact, if we continue the series beyond the segment shown in

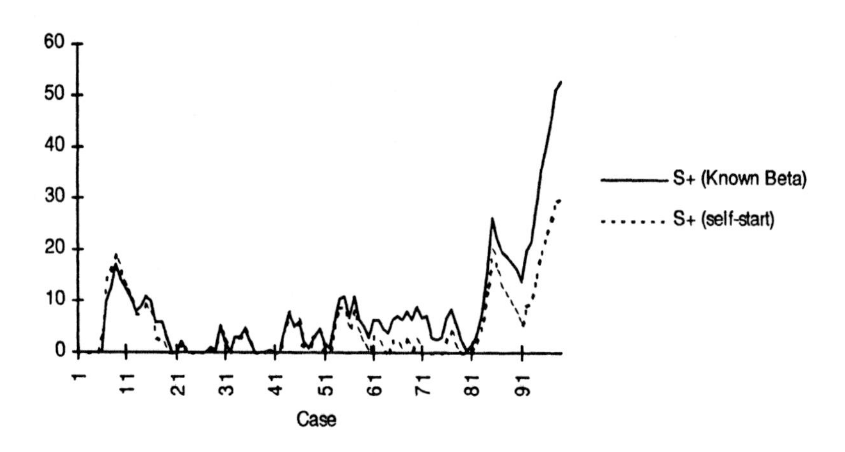

FIGURE 7.2. Known-parameter and self-starting gamma CUSUMs

Figure 7.2, the self-starting CUSUM turns downward and falls below the decision interval again.

As with the normal location CUSUM, the self-starting CUSUM for variance has done its job by signaling the shift. It is up to the user to be alert and react to the signal, and not suppose that because the CUSUM had come back below the decision interval the underlying problem had righted itself without any corrective action.

As with the self-starting CUSUM for normal means, that for gamma-distributed data can also be used to get a Shewhart chart. We illustrate this with the same data set. The transforms U_n are χ_4^2 and so their square roots behave like sample standard deviations of rational groups of size 5 whose true standard deviation is 2. Their true mean is given by standard S chart results as $2c_4 = 1.88$ and their true standard deviation as 0.47. So a conventional Shewhart S chart would have center line 1.88 and upper control limit $1.88 + 3 \times 0.47 = 3.29$. Thus a case whose $\sqrt{U_n}$ was above 3.29 would give a Shewhart signal.

For the sample data set, this is the case with reading number 8. Its U_n of 16.37 corresponds to a standard deviation of 4.05, which exceeds the control limit.

This type of combined Shewhart-CUSUM self-starting analysis can provide a simultaneous control for both isolated and persistent shifts in the scale parameter of gamma data.

7.3.4 Normal data — control of mean and variance

We started this chapter out with a discussion of a self-starting CUSUM of the mean of normal data, and we only covered individual observations. When discussing control of variances, we implied using rational groups. This leaves two logical gaps — how do we handle scale charting for individuals; and how do we control for the means of rational groups?

Beginning with the first of these, with individuals we use the distributional model $X_n \sim N(\mu, \sigma^2)$. The deviations $X_n - \overline{X}_{n-1}$ provide the basis for control on the location, and their squares $(X_n - \overline{X}_{n-1})^2$ provide the basis for control on the scale. The $(X_n - \overline{X}_{n-1})^2$ follow independent scaled $\sigma^2\chi^2$ distributions, so we could go through the same procedure we used for generic gamma problems to divide each by a running estimate of σ^2, get its tail area under its in-control F distribution, and then convert this to an equivalent χ_1^2 variate.

It turns out when you do this, though, that the equivalent χ_1^2 variate is just the square of the equivalent standard normal U_n that you set up for the CUSUM of the mean.

So the combined CUSUM scheme for location and scale consists of using the sequence of running means $\overline{X}_n$ and standard deviations s_n, to form the quantities

$$T_n = \frac{X_n - \overline{X}_{n-1}}{s_{n-1}}$$

converting them to the independent $N(0, 1)$ quantities

$$U_n = \Phi^{-1}[F_{n-2}(a_n T_n)]; \qquad a_n = \sqrt{\frac{n-1}{n}},$$

and forming one CUSUM of the U_n for location, and a separate CUSUM of the χ_1^2 quantities U_n^2 for scale. We remind you that these two CUSUMs are not entirely independent; scale changes will affect not only the scale CUSUM, but also the location CUSUM, and shifts in mean will affect not only the location CUSUM, but also the scale CUSUM. The first of these characteristics is always true (scale changes affect the ARL of the location CUSUM), and the second comes about because we center each X_n using the running mean, and so if there is a location shift it will cause the scale CUSUM to drift upward.

These problems of "bleeding" from one CUSUM into the other are not really serious; although both shifts will ultimately signal on both charts, location shifts tend to signal on the location CUSUM long before they do on the scale CUSUM and vice versa.

If we do have rational groups of size bigger than 1, then it is possible to get the scale chart clear of the effects of shifts in the mean. Write X_n for the mean and, say V_n for the variance of the m observations in the n rational group.

The self-starting scale CUSUM of the V_n is exactly as we outlined it in Section 7.2. The self-starting CUSUM of the mean needs a little adaptation from that of Section 7.1. Let $\overline{X}_n$ be the running mean of the first n rational group means X_n, and write the sequence of running estimates of σ,

$$s_n = \sqrt{\Sigma_1^n V_i / n},$$

then

$$X_n - \overline{X}_{n-1} \sim N\left(0, \frac{n}{m(n-1)}\sigma^2\right).$$

So if we form

$$T_n = a_n(X_n - \overline{X}_{n-1})/s_{n-1}; \quad a_n = \sqrt{\frac{m(n-1)}{n}},$$

then the sequence T_n will be independent, having Student's t distribution with $(n-1)(m-1)$ degrees of freedom. This is all we need to convert the T_n into a stream of equivalent $N(0,1)$ quantities for the location CUSUM.

In this CUSUM pair, scale shifts will still affect the ARL of the location CUSUM (this effect is unavoidable), but location shifts will not affect the scale CUSUM, since this is based on centering the data of each rational group by its own mean.

7.3.5 Comments

While self-starting CUSUMs can also be defined for the other standard continuous distributions in the exponential family — particularly the inverse Gaussian (Olwell, 1996) — we do not explore this point further here.

In both the cases we discussed in some detail, we transformed our sample-standardized quantity to another that would follow the exact distribution of the known-parameter pivotal. There are other possible transformations. One that comes to mind is to use the exact distribution of the sample-standardized process measure to get a probability, and then to convert this probability to a standard normal score. This normal score could then be CUSUMmed like any other standard normal quantity. The approach we outlined for the gamma distribution seems to us preferable to this CUSUM of equivalent normal scores. We know that in any member of the exponential family the optimal test for a step change from one parameter value to another is a decision interval CUSUM of the natural sufficient statistic of that distribution. A CUSUM of a different function of the natural sufficient statistic would by implication be less effective. The approach we have used has the virtue that, as the history length increases, the quantity CUSUMmed converges to the sufficient statistic, and so will share many good properties of this optimal CUSUM, at least asymptotically. This is lost if we transform to some other quantity, such as a standard normal score.

7.4 Discrete data

The commonly used discrete distributions in the exponential family are the Poisson and the Binomial distribution. The self-starting methodology we outlined for continuous distributions — the normal, gamma and inverse Gaussian — cannot be applied directly to these distributions, or indeed to *any* discrete distribution. The basic reason for this is that discreteness of X_n means that the area A_n can only take on a certain discrete set of values, rather than being uniformly distributed over (0,1) as is the case with the continuous quantities. Furthermore, in the discrete case there are no parameter-free standardized forms analogous to the T_n and U_n of the normal and gamma distributions.

Still, it is possible to make progress toward self-starting approaches, and we sketch two methods. Neither of them is completely satisfactory though, and so the area of self-starting control of discrete data remains only partially resolved.

7.4.1 The Poisson distribution

Suppose the process measure is a count X_n that follows a Poisson distribution with an in-control mean μ_0, now assumed unknown. The complete sufficient statistic for estimating μ_0 from a sample of n in-control readings is the sample mean $\overline{X}_n$.

In the continuous case, we found a pivotal quantity involving X_n and the sufficient statistic, and this pivotal had a distribution that did not involve the unknown parameter. The discrete-case analog of this distribution-free pivotal comes from conditioning. Conditional on the value of $W_n = n\overline{X}_n$, while the process is in control

$$X_n \sim Bi\left(W_n, \frac{1}{n}\right),$$

where the notation $Bi(N, \pi)$ is a shorthand for the binomial distribution with n trials and probability of success on any trial equal to π. This distribution is parameter-free; it does not involve the unknown mean μ_0 or any other unknown quantity. If the mean shifts from μ_0 to μ_1, then the conditional distribution of X_n becomes binomial with a probability

$$\frac{\mu_1}{(n-1)\mu_0 + \mu_1}$$

so that a change in μ will change the binomial probability, upward if $\mu_1 > \mu_0$ and downward if $\mu_1 < \mu_0$. Thus monitoring the binomial probability for upward change will provide a control on whether μ has changed upward, and similarly for downward change.

As in the continuous case, we use this conditional distribution of X_n to get a cumulative probability:

$$A_n = Pr\left[Bi\left(W_n, \frac{1}{n}\right) \leq X_n\right].$$

The big difference from the continuous case is that this cumulative probability is only capable of taking on a limited number of distinct values, since conditioned on W_n, X_n can only assume the values $0, 1, \ldots W_n$.

Because of Basu's lemma, however, the successive A_n are distributed independently.

The two approaches we sketch both involve converting A_n to an equivalent standardized form.

It is possible for A_n to equal 1.0, as in the case where $X_1 = 0$ and $X_2 = 1$. Then $A_2 = 1.0$. More generally, if the sequence starts with several successive zeroes, then A_n is undefined for the initial zeroes, and equals 1 for the first non-zero value. This requires some care in our development of methods to transform A_n.

Transformation to a normal score

This approach (due to Quesenberry (1995)) converts A_n to an equivalent normal deviate

$$U_n = \Phi^{-1}(A_n)$$

which may then be used in a CUSUM chart.

Even though U_n may look like a standard normal quantity, it does not follow a standard normal distribution. This is because the area A_n of which it is a transform is not uniformly distributed over [0,1] (as it was in the continuous examples considered earlier) but takes on one of only W_n distinct values. This "graininess" in the attainable values of U_n is a problem when μ_0 is small, and will lead to U_n having a distribution quite different than the standard normal, which in turn can lead to the ARL being much different than normal-based calculations would suggest.

A second difficulty is that $\Phi^{-1}(1)$ is undefined, yet $A_n = 1$ can occur. This can be addressed by winsorizing A_n.

Transformation to a standard Poisson score

The second approach assumes that we have at least some idea of the value of μ_0. This will almost always be the case, as even short-run and startup processes will have some degree of history behind them. The method will work better the closer our guess is to the truth. We have already mentioned that, although self-starting CUSUMs can in principle be started up with no more than one or two preliminary values, in practice it is better to have at least a few more than this theoretical minimum; that being so, the preliminary guess might be the average of 10 or 20 preliminary readings. In

n	X_n	W_n	A_n	U_n	Y_n
1	4	4	—	—	—
2	2	6	.344	-.402	3
3	3	9	.650	.386	5
4	6	15	.943	1.584	8
5	7	22	.944	1.588	8
6	6	28	.827	.941	7
7	9	37	.969	1.864	9
8	6	43	.712	.561	6
9	10	53	.971	1.891	9
10	5	58	.471	-.072	4

TABLE 7.4. Sample calculations of a self-starting Poisson CUSUM.

the worst case, if it turns out that our initial guess was far from the truth (a fact that will emerge in time as data are gathered from the process), then we can go back and redo the calculations using a different value.

Suppose the "educated guess" at μ is m. We transform X_n to a Poisson variate Y_n with parameter m. This is the value Y_n minimizing

$$\left| \sum_{j=0}^{Y_n} \frac{e^{-m} m^j}{j!} - A_n \right| . \tag{7.3}$$

This transformation is intended to get a Y_n that is truly Poisson with mean m, but because of the graininess in the mesh of attainable A_n values it cannot be entirely successful in this.

Again, if $A_n = 1$ we need a alternate method to select Y_n, as there is no Y_n that minimizes Equation 7.3 when $A_n = 1$. Setting $Y_n = X_n$ is the easiest fix.

We then form CUSUMs of the Y_n to check for changes in μ. An upward shift in μ will slant the A_n toward higher values and so lead to an increase in the Y_n.

We prefer this approach to one using CUSUMs of the normal scores U_n because it uses a summand that has the right asymptotic form. If the guessed value of μ_0 was exactly correct, then as n increased, the transforms Y_n would approach the observed values X_n, so yielding the theoretically correct CUSUM, but without the problem of the unknown true parameter value.

Table 7.4 illustrates the calculations involved in this self-starting Poisson CUSUM. We think the true in-control mean is around 5, and so we set up the transformation target as Poisson with mean $m = 5$. To see how the calculations go, consider the line in Table 7.4, $n = 5$, $X_5 = 7, W_5 = 22$. Evaluating the cumulative distribution of a binomial with 22 trials and probability of success $1/5 = 0.2$ at argument $x = 7$ gives a probability of $A_n = 0.944$. The inverse normal integral at this probability gives $U_5 =$

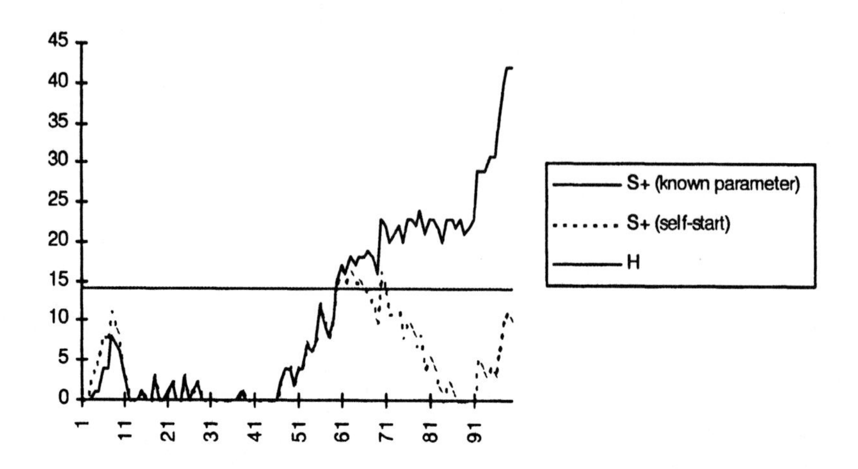

FIGURE 7.3. Known-parameter and self-starting Poisson CUSUMs.

$\Phi^{-1}(0.944) = 1.588$. Using a table of cumulative probabilities of the Poisson distribution with mean $m = 5$, we find that the value with cumulative probability closest to 0.944 is $Y_n = 8$. Thus the process reading X_n can be transformed to an equivalent normal score of 1.588, or to an equivalent Poisson score of 8.

Before leaving the equivalent Poisson score Y_n, we note that it can be used in a Shewhart c chart. By design, the Y_n have a true mean of m ($m = 5$ in this case), and so the c chart for the Y_n would have a center line of $m = 5$ and upper control limit of $m + 3\sqrt{m} = 11.7$. So a Poisson equivalent of 12 or more would signal an isolated special cause on this self-starting Shewhart Poisson chart. None of the cases in our short sequence exceeded this cutoff value.

The values in Table 7.4 are the first 10 cases of a Poisson-distributed series of length 100. The first 50 cases have true mean 5 and the rest have true mean 7. In Figure 7.3, we see a decision interval CUSUM for the self-starting CUSUM of the whole sequence, along with the idealized CUSUM for Poisson data that we could make if we knew the true mean. As the figure shows, the two CUSUM match very well for the first 70 cases. In particular, both CUSUMs signal the shift in mean at the same time. At the end of the series the known-parameter CUSUM is still continuing up about linearly, but the self-starting CUSUM is curving back down to the axis. This behavior is by now familiar. Once the self-starting CUSUM has signaled the shift in the process, it is vital to act on the signal and stop accumulating contaminated information into the running mean. If this is

μ	Y_n	U_n	modified U_n	X_n
5.0	913	180	914	916
5.5	194	68	531	150
6.0	62	27	175	45
6.5	22	14	38	21
7.0	13	10	16	13

TABLE 7.5. ARLs of three Poisson CUSUMs.

not done, then in time the running mean will adapt to the new level and the CUSUM signal will go away.

We have sketched two self-starting approaches: one using a transformation to a known-parameter Poisson variable and one aiming for standard normal scores. Because of the graininess of the A_n, the U_n are not in fact standard normal, but it is not hard to calculate the asymptotic distribution of the U_n, whose true mean and variance turn out to be 0.23 and 0.94 respectively. This means that the in-control mean of the supposedly standard normal quantities is actually a quarter standard deviation off center. This may seem quite small, but it has a profound effect on the in-control ARL. To show this, we did a small simulation, simulating 1,000 runs of Poisson data, first in control and then with various increased means. We used the schemes $k = 6$, $h = 14$ for the two Poisson CUSUMs, and $k = 0.5$, $h = 4.99$ for the normal. These should both give in-control ARLs of 922. The actual simulated ARLs are shown in Table 7.5.

As the table shows, the in-control ARL of the U_n is far below the nominal figure of 922. The actual ARL of 180 means that false alarms would occur at a rate 5 times higher than anticipated. This is a deficiency of the inverse normal transformation.

The in-control ARL of Y_n is fine; it is indistinguishable from that of X_n. Going to the out-of-control scenarios, the ARL of the self-starting CUSUM of Y_n is not much greater than that using X_n and the (usually unavailable) exact information on the process mean. This is most encouraging, indicating that the price paid for not knowing the true mean need not be excessive. So from the viewpoints of both in-control and out-of-control behavior, the CUSUMS of the equivalent Poisson scores Y_n look very satisfactory.

We could get the in-control ARL of U_n right by changing its k and/or h parameters. For example, motivated by the observation that the true in-control mean of U_n is about 0.2 rather than 0, we might correct for this offset by increasing k by 0.2 and leaving h alone. Doing so gives the column headed "modified U_n." As the first row shows, this is successful in getting the in-control ARL right. The ARL of the modified U_n is still poor though. In all out-of-control cases it is well above that of the Y_n. This shows that, in this example at least, the use of normal scores in the self-starting CUSUM gives performance much inferior to that obtained using Poisson scores. This gives some empiric support for the theoretical prediction that transforming to Poisson scores should be more effective than using normal scores since asymptotically the Poisson scores will correspond to the optimal test.

Example

We close this discussion with a data set discussed by Yashchin (1995). The data consist of the numbers of contaminating particles on semiconductor wafers. Experience is that this number has a distribution that changes abruptly from time to time. A methodology for detecting shifts from some unknown mean to some other unknown mean would be useful in adjusting the wafer washing operation.

The usual mean λ is thought to be around 6. We therefore set up a self-starting scheme aimed at a mean of 6 for the transformed values Y_n. We set up upward and downward CUSUMs with reference values $k = 7$ and $k = 5$, respectively. The resulting data are given in Table 7.6. The decision interval for the upward and downward CUSUMs are set to 14 and 12, giving in-control ARLs of 474 and 531, respectively.

At reading 32, the upward CUSUM crosses its decision interval, indicating that there has been an increase in mean. The last in-control reading is estimated to be reading 20. Observations 21 through 32 have accumulated a total of $W_{32} - W_{20} = 224 - 114 = 110$ particles, giving an estimate of $110/12 = 9.1667$ for the current mean.

For continued control, we need to restart the CUSUM. This is done from observation 21. At this stage we have the opportunity to make a fresh choice of the target λ or to stick with the standard value of $\lambda = 6$. Opting for simplicity at some cost in performance, we will stick with $\lambda = 6$. The restarted CUSUM gives the figures in Table 7.7.

This table shows a sharp decrease in mean. It is detected at reading number 45, and the last in-control point is estimated to be reading 41. The segment from readings 42 through 45 has accumulated $224 - 209 = 15$ particles for an average of 3.75 particles per wafer.

The first shift in mean is detected 8 observations after it is thought to have occurred, and the second 3 observations after its occurrence. This reaction is encouragingly fast.

7.4.2 The binomial distribution

The broad framework of the binomial distribution is very like the Poisson: the ideas developed for the Poisson require change of just two distributions to deal with the binomial distribution. The Poisson and binomial distributions that we use for Poisson data are replaced by the binomial and hypergeometric distributions, respectively. Otherwise, the same considerations, observations, and limitations hold for binomial monitoring as for Poisson.

n	X_n	W_n	A_n	Y_n	S_n^+	S_n^-
2	4	11	.274	4	.00	-1.00
3	9	20	.908	9	2.00	.00
4	9	29	.834	8	3.00	.00
5	2	31	.037	1	.00	-4.00
6	10	41	.932	9	2.00	.00
7	3	44	.109	3	.00	-2.00
8	6	50	.564	6	.00	-1.00
9	6	56	.569	6	.00	.00
10	5	61	.421	5	.00	.00
11	5	66	.438	5	.00	.00
12	7	73	.738	7	.00	.00
13	5	78	.439	5	.00	.00
14	7	85	.739	7	.00	.00
15	3	88	.154	3	.00	-2.00
16	4	92	.312	4	.00	-3.00
17	8	100	.866	8	1.00	.00
18	4	104	.309	4	.00	-1.00
19	5	109	.485	5	.00	-1.00
20	5	114	.492	5	.00	-1.00
21	9	123	.930	9	2.00	.00
22	8	131	.857	8	3.00	.00
23	8	139	.847	8	4.00	.00
24	8	147	.838	8	5.00	.00
25	13	160	.995	13	11.00	.00
26	10	170	.935	9	13.00	.00
27	6	176	.522	5	11.00	.00
28	10	186	.929	9	13.00	.00
29	11	197	.958	10	16.00	.00
30	3	200	.097	2	11.00	-3.00
31	11	211	.958	10	14.00	.00
32	13	224	.989	12	19.00	.00

TABLE 7.6. Calculations for self-starting Poisson CUSUM of the wafer data.

n	X_n	W_n	A_n	Y_n	S_n^+	S_n^-
22	8	17	.500	5	.00	.00
23	8	25	.538	6	.00	.00
24	8	33	.553	6	.00	.00
25	13	46	.938	10	3.00	.00
26	10	56	.674	6	2.00	.00
27	6	62	.199	3	.00	-2.00
28	10	72	.714	7	.00	.00
29	11	83	.792	7	.00	.00
30	3	86	.023	1	.00	-4.00
31	11	97	.830	8	1.00	-1.00
32	13	110	.926	9	3.00	.00
33	9	119	.566	6	2.00	.00
34	11	130	.781	7	2.00	.00
35	13	143	.904	9	4.00	.00
36	15	158	.961	10	7.00	.00
37	6	164	.146	3	3.00	-2.00
38	10	174	.626	6	2.00	-1.00
39	11	185	.730	7	2.00	.00
40	12	197	.811	8	3.00	.00
41	12	209	.801	8	4.00	.00
42	2	211	.003	0	.00	-5.00
43	4	215	.041	2	.00	-8.00
44	7	222	.290	4	.00	-9.00
45	2	224	.006	0	.00	-14.00

TABLE 7.7. Calculations for the restarted self-starting Poisson CUSUM of the wafer data.

If the measurement at each sampling occasion is the number X_n of nonconforming items in a rational group of size m_n, then the basic in-control model for X_n would be a binomial distribution with m_n trials and probability π. We can summarize the information about π from the first n in-control samples with the two summary statistics:

$$N_n = \sum_{j=1}^{n} m_j$$
$$W_n = \sum_{j=1}^{n} X_j$$

(the running total sample size and total number of successes). If we condition on the total number of nonconforming items W_n seen among the N_n sampled units in the first n samples, then the conditional distribution of X_n is hypergeometric. (It is like an acceptance sampling problem in which we have a lot with W_n nonconforming items, $N_n - W_n$ conforming, and X_n is the number of nonconforming items in a sample of size m_n.) Its probability distribution is

$$p_n(x) = \frac{\binom{N_n - W_n}{m_n - x}\binom{W_n}{x}}{\binom{N_n}{m_n}}.$$

With this result we can compute each X_n's cumulative probability A_n under this hypergeometric distribution.

$$A_n = \sum_{x=0}^{X_n} p_n(x).$$

The sequence of A_n will be independent, thanks once again to Basu's lemma. They will not be exactly uniformly distributed, because X_n can take on only a finite number of different values.

As with the Poisson self-starting CUSUM, we must pay attention to the special case where our leading observations are $X_1 = 0, X_2 = 0, \ldots, X_{k-1} = 1, X_k = i$, where $i > 0$. Then $W_k = X_k$, and $A_k = 1$.

The same two uses come to mind for these A_n. One is to convert them to standard normal scores

$$U_n = \Phi^{-1}[A_n]$$

an approach again discussed in more detail in Quesenberry (1995). As the A_n are not uniformly distributed but have the same sort of graininess we get with Poisson data, these U_n will not be exactly normal; each U_n has a limited range of possible values and, as we saw with the Poisson case, this graininess might have a large impact on the ARL of CUSUMs of the U_n. And, of course, the case where $A_n = 1$ requires some attention.

The other intuitively reasonable approach is to convert the A_n to equivalent standard binomial quantities Y_n following a binomial distribution with m items sampled and probability of nonconforming p. We do this by solving for Y_n as closely as possible

$$A_n = \sum_{j=0}^{Y_n} \binom{m}{j} p^j (1-p)^{m-j}.$$

If $W_n = 0$, we set $Y_n = 0$.

Ideally, we would like p to be close to π, and often process knowledge will give us a usable guess for a π to use. It is also convenient to use a fixed value for the equivalent sample size m, even if the sample sizes of the actual data m_n are not themselves constant. The m value of the target should be a reasonable substitute for the actual m_n if we are to get the full benefit of the asymptotic optimality, so using a fixed m for the equivalent binomial scores is likely to work best if the m_n do not vary dramatically.

As in the Poisson case, the target Y_n will not be exactly binomial, because of the graininess in the A_n and because of variation (if any) in the m_n, but should be close to binomial.

The theoretical attraction of this approach is that in the constant m_n case, if our target probability p were close to the true probability π, then as the length of in-control history increased, the transform Y_n would tend more and more to have the value X_n. Since the CUSUM of X_n is the optimal procedure for detecting a shift in π, this suggests that as the sample size grew, the CUSUM of Y_n would attain close to this best-possible performance. Using any other transform — for example, to standard normal scores — would not have this large-sample optimality justification.

Example

We illustrate the self-starting binomial CUSUM with the sentence data introduced in Chapter 5. Each observation consists of the length m_n of a sentence in a computer manual, and X_n the number of words of 3 or more syllables in that sentence. A quick scan of the data suggests that the sentence length varies around 18 words, and that about 15% of the words are of 3 or more syllables. We therefore use a binomial with $m = 18, p = 0.15$ as the target distribution.

Table 7.8 illustrates the calculations of the Y_n and the upward and downward CUSUMs. For the choice $m = 18, p = 0.15$, the mean is 2.7 long words per sentence; we set up the reference values $k = 3.5$ for an upward CUSUM and $k = 2$ for a downward CUSUM.

A graph of the CUSUMs is shown in Figure 7.4. The choices $h^+ = 7.5$ and $h^- = 7$ correspond to in-control ARLs of 529 and 696, respectively, and are shown in the figure. The self-starting CUSUM broke through the upward decision interval shortly before reading 50. Reading the input data

n	m_n	X_n	N_n	W_n	A_n	Y_n	S_n^+	S_n^-
2	20	4	30	5	.891	4	.50	.00
3	14	5	44	10	.961	5	2.00	.00
4	14	1	58	11	.186	1	.00	-1.00
5	22	0	80	11	.022	0	.00	-3.00
6	13	1	93	12	.472	2	.00	-3.00
7	22	4	115	16	.839	4	.50	-1.00
8	33	4	148	20	.525	2	.00	-1.00
9	23	1	171	21	.186	1	.00	-2.00
10	27	1	198	22	.161	1	.00	-3.00
11	30	6	228	28	.946	5	1.50	.00
12	22	4	250	32	.868	4	2.00	.00
13	23	4	273	36	.831	4	2.50	.00
14	13	2	286	38	.759	3	2.00	.00
15	7	0	293	38	.374	2	.50	.00
16	15	1	308	39	.411	2	.00	.00
17	34	4	342	43	.570	2	.00	.00
18	9	1	351	44	.686	3	.00	.00
19	8	2	359	46	.931	5	1.50	.00
20	19	3	378	49	.778	3	1.00	.00

TABLE 7.8. Calculations for a self-starting CUSUM of sentence length data.

to the CUSUM shows that the signal occurs at reading 47, with reading 36 indicated as the last in-control case. At reading 36, $N_{36} = 695$, $W_{36} = 80$, so for these first 36 sentences the average proportion of long words is 0.12. $N_{47} = 895$ and $W_{47} = 119$, so the sentences 37 through 47 have added 200 words, 39 of them long, for an average proportion of long words of 0.33.

It seems that, having started out with short words on the introductory material, the book has run into a patch with a much higher proportion of long words.

The upward CUSUM then comes down below the decision interval. We should not make too much of this; the patch of long words will have increased the average proportion of long words and so "moved the goalpost" higher. To properly diagnose further changes from this new higher-level mean we should restart the procedure from reading number 37.

7.4.3 Updating the targets

The self-starting CUSUMs for Poisson and binomial data had an element that was not present in the self-starting CUSUMs of continuous data: it was necessary to specify a target mean to which the data would be transformed. The method will presumably work best when the target mean is close to the true in-control mean, and presumably could be really bad if the target mean is very far from the true in-control mean.

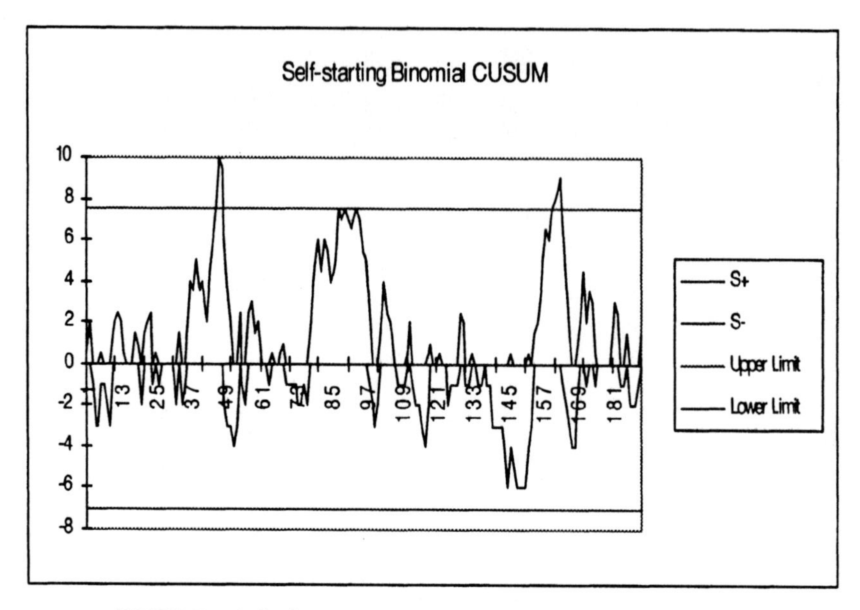

FIGURE 7.4. Self-starting binomial CUSUMs of sentence data.

Although there is no reason to think that a good guess at the true mean will not work well, we can periodically respecify the target mean by using the data accumulated up to that point to give an updated figure for the target mean. We could do this mechanically, by using the running mean directly as the value of m (Poisson) or mp (binomial), or we could use the mean in a more Bayesian way to modify our prior belief about the in-control true mean.

We repeat that this updating is not essential from a theoretical perspective, but is likely to enhance the behavior of the chart.

7.5 Summary

- Self-starting CUSUM charts are a "no lose" proposition.

- Transforming discrete data to normal scores for CUSUM purposes is inefficient.

- Schemes based on distributions which asymptotically converge to the known-parameter case behave well.

7.6 Further reading

Bagshaw and Johnson (1975) show in more detail than our little 3×3 table the impact of misspecification of σ on CUSUM behavior.

Short run processes have been the object of a considerable literature, though the particular recursive approach we use in the self-starting CUSUMs has not. With short runs, it may be more attractive to use change-point formulations which test for a shift at some unknown instant from one unknown set of parameter values to another unknown set of parameter values – two useful tests are for example Worsley (1979) for a change-point formulation for a step change in a normal mean with unknown variance and Worsley (1986) for a step change in exponential data. Srivastava and Worsley (1986) covers a multivariate test which is related to the material in Chapter 8.

The change-point formulation has the attraction of greater statistical efficiency if there is indeed a step change from one level to another, but the drawback that if the change were other than a step change, detection could be less effective than that provided by the fairly sturdy self-starting CUSUM.

We mentioned the statistical underpinning of the exact self-starting CUSUM lies in Basu's theorems on independence of pivotals of complete sufficient statistics. The most familiar application of this theory may be that to "recursive residuals" in linear multiple regression. We will say more about recursive residuals in Chapter 9.

Olwell (1997a, 1997b) has explored the Bayesian approach to self-starting charts in greater detail.

8
Multivariate data

> Generally, management of many is the same as management of few. It is a matter of organization.
>
> *Sun Tzu*

Introduction

Many processes involve multiple process measurements. These can be of several types, as the following examples illustrate.

- Different properties may be measured on each unit produced. In manufacturing roller bearings, for example, we might measure the length, maximum diameter, and minimum diameter of each sampled bearing.

- In process control, we often see a number of different but connected processes. The measurements made on the different processes may then be related to each other. For example, in semiconductor wafer fabrication, chips go through sequences of processing steps. The quality of a chip at one stage depends, not only on this most recent process step performed on it, but also on the outcomes of all the earlier processing steps. To look at the measures on one step in isolation from those on previous steps then is a bad idea. The product at some stage may be bad, not because of problems in that stage, but because of problems created in earlier stages. Not recognizing this possibility can lead to tinkering with a process that is working perfectly well, which wastes time and resources, and is also likely to make quality worse rather than better.

- A related problem has some of the process steps located outside the operation being controlled. For example, in a coal washing plant, the yield and ash content of the washed coal are important measures of the quality of the product and the effectiveness of the washing plant. Both however are strongly affected by the quality of the raw coal entering the plant, and this is something outside the control of the washing plant. Still the plant should measure these properties of the incoming coal so that the yield and washed ash figures can be "adjusted for" the quality of the coal supplied. Making the adjustment leads to more sensitive SPC controls and to more accurate recognition of when there are and are not process problems.

The basic framework of this chapter is not a single process measure at each time point (as we have had up to now), but a *vector* of different but possibly related measures.

There are two ways to think about a vector of process measures like this. One is to ignore the connection between the different measures and treat the problem as a collection of unrelated univariate problems. This is what happens by default if you do not consciously study the related items as a whole.

The other way is to think about the collection of measures as a multivariate measure and control it as such. Doing this potentially brings several benefits. A major benefit is that multivariate control may be much more sensitive to shifts than is the collection of univariate control, but this is not the only benefit. Another is that multivariate control can be more specific in diagnosing causes than is a collection of univariate charts. The example of the coal washing plant illustrates this: a sudden increase in the ash content of the washed coal could be attributed to problems in the washing when in fact it was due to a deterioration in the quality of the incoming coal. Not recognizing the true source of the problem would then lead to mistakenly looking for a non-existent problem in the washing plant.

Multivariate control can resolve this error.

- Multivariate control offers increased sensitivity to departures from in—control.
- Multivariate control offers improved diagnostics.

8.1 Outline of the multivariate normal

The univariate CUSUM was first introduced within the framework of a normal distribution, not necessarily because all process measures follow normal distributions, but because the normal distribution is well known, has "nice" statistical properties, and can often be fitted to transformed data even if it does not fit the original measurements all that well.

In multivariate data, the normal distribution is even more central, as there are few other manageable widely known distributions available. For this main reason, the multivariate normal distribution will be the focus of the discussion of multiple process measures.

In this chapter, we assume familiarity with the multivariate normal distribution to the level of standard introductory texts on multivariate analysis. The statistical level is necessarily higher than it is for the other material in this book.

Write p for the number of related measurements taken and $\boldsymbol{X}_n$ for the p-component process measurement vector made on the nth sample. The multivariate normality assumption then is that the vectors $\boldsymbol{X}_n$ follow a common multivariate normal distribution with some mean vector $\boldsymbol{\mu}$ and some covariance matrix $\boldsymbol{\Sigma}$. We abbreviate this to $\boldsymbol{X}_n \sim \boldsymbol{N}(\boldsymbol{\mu}, \boldsymbol{\Sigma})$.

The covariance matrix $\boldsymbol{\Sigma}$ captures the strength of the relationship between the different process measures made on the same sample. If its off-diagonal elements are all zero, then the process measures are uncorrelated, and there may be little or no benefit in thinking about the process control as a multivariate problem rather than a collection of univariate problems. We do need to say *may be* rather than *is*, as multivariate control can be better than a collection of univariate charts even if the attributes are unrelated if special causes tend to affect several of the properties measured.

The model we are using assumes that the vectors $\boldsymbol{X}_n$ are independent for different n. Thus although the different measures on the same sample may be correlated, the measures at one sampling are independent of those at another sampling.

A comment on defining the vector $\boldsymbol{X}_n$ may be useful here. We need to define the different measures so that they may be correlated with other measures in the same measurement vector, but not with any measurements in any other measurement vector. To illustrate the relevance of this, consider, say, a two-step batch chemical synthesis in which a source material is processed for 30 minutes in one vessel and an intermediate product then goes to another vessel where it is processed for another 30 minutes. Process measurements are made on the raw material going into the first vessel, on the intermediate product coming out of the first vessel, and on the final product taken from the second vessel. When assembling the three process measures into a single vector, it is important to match the three measures so that they refer to the same material. So, for example, if a sample is taken of each material every half hour, it would be wrong to construct a measurement vector by joining up the raw, intermediate, and final products measured at the same time. This is because these measures would refer to different batches. Instead, the raw material observations made at 11:00 should be joined to the intermediate material observation made at 11:30 and the final product observation made at 12:00. This time matching would correctly reflect the process matching.

In the context of a continuous batch process like this, the point about taking care to match the measures appropriately is perhaps too obvious to need much discussion, but in other contexts the need to match the measures may be equally great but subtle enough to be missed.

> The multivariate normal distribution is our model of choice.

8.2 Shewhart charting — Hotelling's T^2

One of the early applications of multivariate analysis was to a multivariate quality control problem. This led Hotelling to use the T^2 test statistic that bears his name. Suppose the p-component vector $\boldsymbol{X}$ follows a multivariate normal $\boldsymbol{X} \sim \boldsymbol{N}(\boldsymbol{\mu}, \boldsymbol{\Sigma})$ with $\boldsymbol{\Sigma}$ known, and we want to test the null hypothesis

$$H_0 : \boldsymbol{\mu} = \boldsymbol{\mu}_0$$

against the alternative hypothesis

$$H_a : \boldsymbol{\mu} \neq \boldsymbol{\mu}_0,$$

where the alternative hypothesis does not specify what value the mean vector $\boldsymbol{\mu}$ might have if it does not have the value $\boldsymbol{\mu}_0$. Since the alternative hypothesis does not specify any particular value for $\boldsymbol{\mu}$, it seems to make sense to restrict attention to *affine invariant* test statistics — test statistics whose value is unaffected by a full-rank linear transformation of the vector $\boldsymbol{X}$. Multivariate theory then shows that the most powerful affine invariant test statistic for H_0 against H_a rejects the null hypothesis if the value of

$$T^2 = (\boldsymbol{X} - \boldsymbol{\mu}_0)'\boldsymbol{\Sigma}^{-1}(\boldsymbol{X} - \boldsymbol{\mu}_0).$$

is sufficiently large. The question of how large is sufficiently large is then answered by the distributional result that

$$T^2 \sim \chi^2_p.$$

This observation then leads to a multivariate quality control application. Suppose that while in control the process measurement vector $\boldsymbol{X}_n$ follows a $\boldsymbol{N}(\boldsymbol{\mu}_0, \boldsymbol{\Sigma})$ distribution, but that when the process is out of control the mean vector shifts to some other unknown value. If we restrict ourselves to affine invariant tests, then the optimal test statistic is Hotelling's T^2

$$T^2 = (\boldsymbol{X}_n - \boldsymbol{\mu}_0)'\boldsymbol{\Sigma}^{-1}(\boldsymbol{X}_n - \boldsymbol{\mu}_0)$$

An extension of this basic result is more useful in practice. If instead of a single vector $\boldsymbol{X}_n$ we take a rational group of size m and compute their sample mean vector $\overline{\boldsymbol{X}}_n$, then

$$\overline{\boldsymbol{X}}_n \sim \boldsymbol{N}(\boldsymbol{\mu}_0, \frac{\boldsymbol{\Sigma}}{m}),$$

giving the optimal test statistic

$$T^2 = m(\overline{\boldsymbol{X}}_n - \boldsymbol{\mu}_0)'\boldsymbol{\Sigma}^{-1}(\overline{\boldsymbol{X}}_n - \boldsymbol{\mu}_0).$$

If the mean vector shifts to some new value, say, $\boldsymbol{\mu}_1 = \boldsymbol{\mu}_0 + \boldsymbol{\Delta}$, then the distribution of T^2 changes from central to noncentral χ^2 with p degrees of freedom and noncentrality

$$\lambda = m\boldsymbol{\Delta}'\boldsymbol{\Sigma}^{-1}\boldsymbol{\Delta} \tag{8.1}$$

(the case of single observations corresponding to $m = 1$).

This general optimality result for T^2 is very far-reaching. It shows that T^2 is a general metric for turning measurement vectors into scalars that retain the essential information about whether the mean vector is indeed $\boldsymbol{\mu}_0$ or some other unspecified vector.

Valuable though this optimality result is, we should not lose sight of the fact that the optimality of T^2 is within the class of affine invariant tests. If we have reason to believe that when $\boldsymbol{\mu}$ shifts it does so in a particular direction, then we can vastly improve on the T^2 test. Suppose, for example, that the alternative hypothesis is

$$H_a : \boldsymbol{\mu} = \boldsymbol{\mu}_1,$$

where $\boldsymbol{\mu}_1$ is specified. Then the optimal test statistic becomes

$$Z = (\boldsymbol{X} - \boldsymbol{\mu}_0)'\boldsymbol{\Sigma}^{-1}(\boldsymbol{\mu}_1 - \boldsymbol{\mu}_0).$$

This follows a (univariate, scalar) normal distribution with λ given by Equation 8.1.

$$Z \sim N(0, \lambda), \qquad \boldsymbol{\mu} = \boldsymbol{\mu}_0;$$
$$Z \sim N(\lambda, \lambda), \qquad \boldsymbol{\mu} = \boldsymbol{\mu}_1.$$

This approach is always more powerful than T^2, and its advantage is greater the larger p is. To see this, if you square Z and divide by λ you get a χ^2 distributed with 1 degree of freedom, and noncentrality (when H_a

is true) of λ. This is the same noncentrality that T^2 has, but as it is on a χ^2 with 1 degree of freedom rather than p, this noncentrality buys more performance. The larger p, the greater the advantage in using Z rather than T^2.

Note that it is not necessary for the alternative hypothesis to specify the actual *value* of $\boldsymbol{\mu}_1$ — just its *direction* from $\boldsymbol{\mu}_0$. Different choices of $\boldsymbol{\mu}_1$ of the form $\boldsymbol{\mu}_1 = \boldsymbol{\mu}_0 + c\Delta$ with the same Δ but different scalar multiples c all lead to the same (apart from trivial rescaling) test statistic Z.

Looking back at the formula for Hotelling's T^2, you can see that T^2 is just this optimal test, but modified by "cheating" to make it look for a shift in the apparent estimated direction of $\boldsymbol{\mu}_1$.

To summarize, if we have no idea of where $\boldsymbol{\mu}$ might go when the process is out of control, then it makes sense to use Hotelling's T^2 because of its optimality among affine invariant tests. At the other end of the spectrum, if there is only one value to which $\boldsymbol{\mu}$ would shift when the process was out of control, then you would get much better performance by using the statistic Z, which is not affine invariant but which incorporates the process knowledge of the expected out-of-control behavior. Many realistic problems lie between these two ends of the spectrum; process knowledge indicates that some directions of shift are much more likely to occur than others. Procedures that incorporate this directional information can then be used for better performance than the T^2 metric gives.

8.3 CUSUM charting — various approaches

Hotelling introduced the T^2 statistic in the context of a multivariate Shewhart style control problem of checking the mean of rational groups for conformity with an in-control multivariate normal distribution. Just as in the univariate case, this is only one half of the problem; the other half is methods that will detect smaller but persistent shifts. The need for some method that accumulates information across successive observations is therefore compelling, but setting up a cumulative scheme for multivariate data is easier said than done. In the univariate case, we defined the decision interval CUSUM by the recursion

$$S_n = \max(0, S_{n-1} + X_n - k)$$

but in the case of multivariate data where the scalar X_n is replaced by a vector $\boldsymbol{X}_n$, it is not clear how we might define something analogous to the max of zero and some update of the previous value of the CUSUM.

In the univariate case, we saw that the decision interval CUSUM was the optimal diagnostic for a shift in a parameter from some in-control value μ_0 to some out-of-control value μ_1. In the case of exponential family parameters, this led to a CUSUM of the sufficient statistic, with the in-control and out-of-control parameters appearing in the problem only in that they define the reference value k. Thus different choices for the out-of-control parameter value μ_1 would lead to a CUSUM using the same summand, but with a different reference value.

The likelihood ratio methodology of Chapter 6 does not really distinguish between a scalar and a vector process measure; the optimal diagnostic for a shift from the in-control mean vector $\boldsymbol{\mu}_0$ to the out-of-control mean vector $\boldsymbol{\mu}_1$ is a CUSUM (suggested by Healy) of the same score we defined earlier for a shift in a known direction — the normally distributed scalar

$$Z_n = (\boldsymbol{X}_n - \boldsymbol{\mu}_0)'\boldsymbol{\Sigma}^{-1}(\boldsymbol{\mu}_1 - \boldsymbol{\mu}_0).$$

The reason this is a less than complete solution to the problem of multivariate CUSUM control is that the out-of-control level $\boldsymbol{\mu}_1$ does not affect something as minor as the reference value, but changes the whole form of the CUSUMmand. If the shift is in some other direction than $\boldsymbol{\mu}_1$, the CUSUM of Z_n could be anything from quite effective to completely useless.

This leads to the question of how to set up multivariate CUSUM-like procedures that will be sensitive to shifts in all directions, rather than just the single direction required by the CUSUM of Z_n.

One thought that occurs naturally is to set up a CUSUM of the Hotelling T^2, perhaps reasoning that, if an individual T^2 is in any sense optimal, then accumulating its value across successive cases should also work well. Intuition is not a good guide here though, and CUSUMs of T^2 and of T are not particularly successful. One deficiency of T^2 is perhaps apparent when we realize that it is an optimal diagnostic for a step change, but not for a step change in the mean vector. It is optimal for a step change in the *covariance matrix* from $\boldsymbol{\Sigma}$ to $a\boldsymbol{\Sigma}$ for a constant $a > 1$. Although the CUSUM of T_n^2 would obviously tend to drift upward if the mean vector shifted, it would presumably be affected much more strongly by overall increases in the variance of the measures, and so have the deficiency of confusing location shifts with scale shifts.

In the scalar normal case, the CUSUM of X_n itself (whose sign can be positive or negative) was able to avoid this difficulty (scale changes make the normal mean CUSUM more variable, but do not give it a systematic drift), but in the multivariate case T^2 is inherently positive, and it is not so easy to see how to accumulate on the $\boldsymbol{X}$ scale.

Crosier (1988) proposed a multivariate CUSUM that does accumulate on the scale of $\boldsymbol{X}$. This initializes the CUSUM vector $\boldsymbol{S}_n$ to a zero vector, and then uses the recursion

$$\boldsymbol{S}_n = \begin{cases} 0; & \text{for } C_n \le k \\ (\boldsymbol{S}_{n-1} + \boldsymbol{X}_n - \boldsymbol{\mu}_0)/(1 - k/C_n); & \text{for } C_n > k \end{cases}$$

and where we define

$$C_n = (\boldsymbol{S}_{n-1} + \boldsymbol{X}_n - \boldsymbol{\mu}_0)'\boldsymbol{\Sigma}^{-1}(\boldsymbol{S}_{n-1} + \boldsymbol{X}_n - \boldsymbol{\mu}_0).$$

The CUSUM signals if $S_n' \Sigma^{-1} S_n > h$, where h is the (scalar) decision interval.

This CUSUM is intuitively attractive in that it is accumulating on the X scale rather than a quadratic scale, it has the same property as the scalar CUSUM of resetting to zero when there seems little evidence that the process is off target; and its final decision uses the T^2 metric. It has no known optimality properties, but does appear to have good practical performance.

8.3.1 Collections of univariate CUSUMs

The final idea that comes to mind in CUSUMming multivariate data is a collection of unconnected univariate CUSUMs of the individual measures. This proposal, the MCUSUM, was put forward by Woodall and Ncube (1985). It formalizes what probably happens quite commonly, that each measure is CUSUMmed individually, and an out-of-control signaled if any of the individual CUSUMs crosses some decision interval. Once we think about this collection of CUSUMs as a single control scheme, it is necessary to adjust the decision interval to reflect the fact that each chart adds its own false alarms to the system, so that to keep the total false alarm rate low it is necessary to reduce the false alarm rate (i.e., increase the in-control ARL) of each of the individual charts.

This approach has the drawback that it does not make any direct use of the correlation between the different measures. For this reason, it may be inferior to methods that do take account of this correlation.

Example: Ambulatory monitoring. An interesting nontraditional application of statistical process control is in the area of "ambulatory monitoring" in which people spend long periods of time wearing equipment that measures and records some physiological variable. We illustrate with a data set (kindly provided by Dr. Franz Halberg of the University of Minnesota with consent by their collector, Dr. Yoshihiko Watanabe of Tokyo) in which the wearer's blood pressure and heart rate were measured and recorded every 15 minutes for 6 years. Prior to analysis using SPC methods, each week's raw data are condensed into 8 summary numbers:

- SBPM — the MESOR of systolic blood pressure;
- DBPM — the MESOR of diastolic blood pressure;
- MAPM — the MESOR of mean arterial pressure;
- HRM — the MESOR of heart rate;
- SBPA — the circadian amplitude of systolic blood pressure;
- DBPA — the circadian amplitude of diastolic blood pressure;

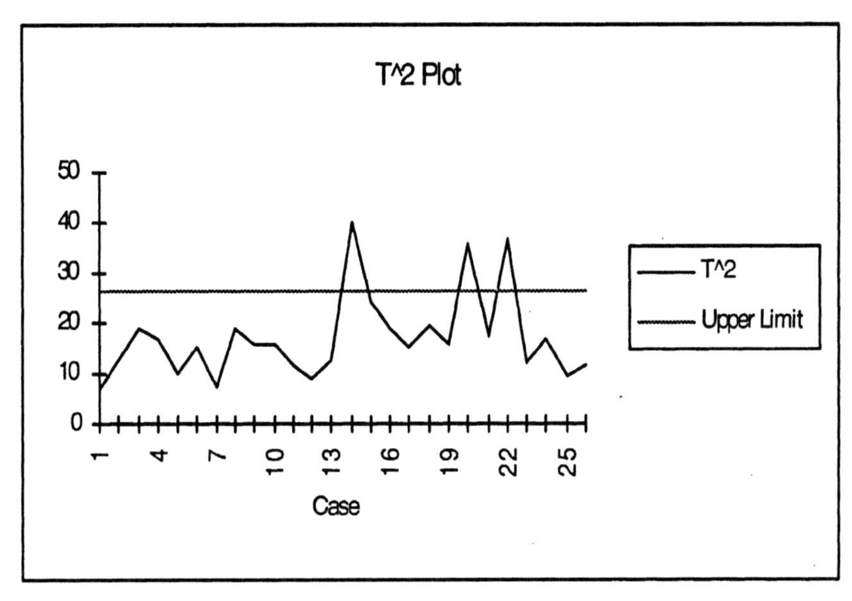

FIGURE 8.1. Control chart of T^2 for the ambulatory data.

- MAPA — the circadian amplitude of mean arterial pressure;
- HRA — the circadian amplitude of heart rate.

Loosely speaking, the MESOR is a location summary statistic and the amplitude measures scale. High mean blood pressure (as reflected in a high MESOR) is well known to be a risk indicator for heart attacks and strokes; it is less widely appreciated that the surges in pressure associated with a high amplitude are also dangerous. SPC procedures applied to these series can identify times at which these characteristics change level, and at which therefore the health risks increase or decrease. In addition to this clinical issue of whether the level of any of the measures has changed, there are also instrumental issues such as malfunction, miscalibration, and drifts in the instrument.

We discarded the first year's data (for reasons that emerge in the following) and used the next 150 data vectors — approximately three years' data — to estimate the mean vector and covariance matrix. A chart of T^2 is given in Figure 8.1. This plot shows that toward the end of the period, several of the vectors were beyond the control limit.

As already mentioned, we are not very enthusiastic about CUSUMs of T^2 for control of the mean vector, but they certainly are easy enough to produce and to calibrate. While the process is in control, T_n^2 follows a χ^2 distribution with p — in this case 8 — degrees of freedom. If we design for, say, a reference value of $k = 12$, then the decision interval $h = 17.12$ gives an in-control ARL of 1,000.

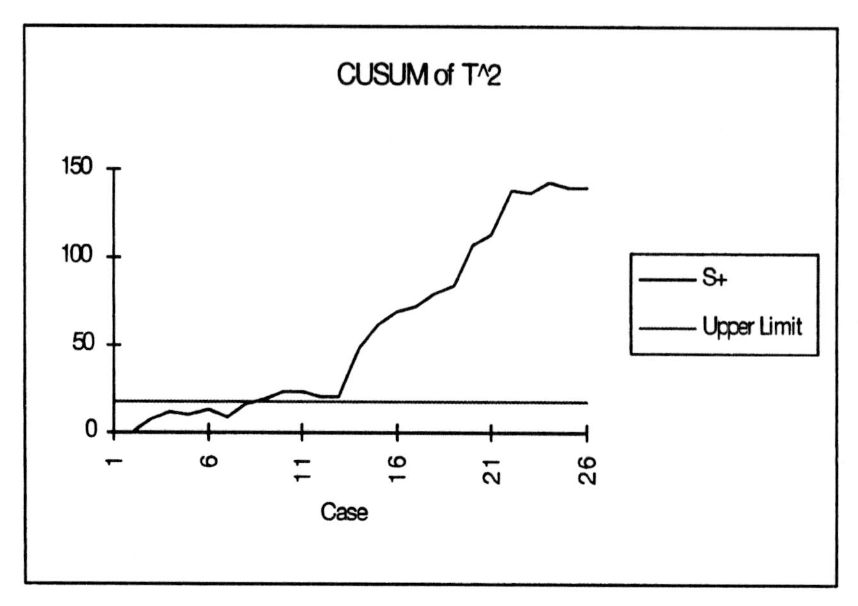

FIGURE 8.2. CUSUM of T^2 for the ambulatory data.

Figure 8.2 shows a CUSUM of the T^2 values, using this reference value and decision interval. This plot shows the change earlier than it appeared on the T^2 chart; at week 9 the CUSUM breaks out of the decision interval.

Finally, we show the same data set put through Crosier's proposal using a reference value of $k = 9$. The chart of the length of $\boldsymbol{S}_n$ is shown in Figure 8.3. This chart broke out of its decision interval at observation 5, and continued to climb right through the rest of the series.

In this data set, all the multivariate charts have been easily able to detect a substantial shift that seems to have happened about the time the monitoring was started.

8.4 Regression adjustment

A major deficiency of methods based on T^2 (a class that includes Crosier's proposal) is that when they signal, they do not provide any direct diagnosis of what it is that went out of control. To follow up on the special cause that led to the signal, we need to consider:

- whether some variable or variables had a mean shift;
- whether the relationship connecting some variables changed;
- whether the random variability increased; or
- some combination of the preceding.

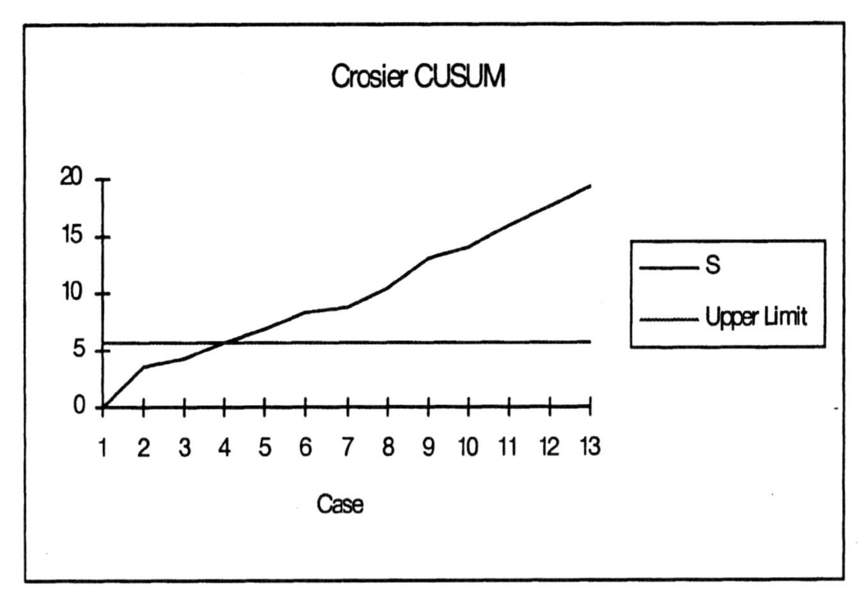

FIGURE 8.3. Crosier CUSUM of T^2 for the ambulatory data.

All of these can lead to increases in T^2.

A useful tool for this sort of follow-up is regression adjustment. This topic is interesting, both in its own right, and because it adds some valuable tools to the multivariate SPC toolkit. Returning to the case of a scalar process measure, suppose that the process measure of interest X is partly driven by a vector of "covariates" $\boldsymbol{W}$. For example, in weaving, the strength of the final fabric is dependent, partly on the processes used in manufacture, but also partly on the properties of the raw fiber used for the weaving. Here X could be the fabric strength, and $\boldsymbol{W}$ a vector of properties of the raw fiber.

One obvious approach to control using X is to measure and chart the X values without regard to the $\boldsymbol{W}$ values. The other approach is to "adjust" the measured X values in some way for the associated $\boldsymbol{W}$ values. With this approach, if the strength of the fabric at any stage drops, but the drop is attributable to a drop in the quality of the raw fiber, then the corrective action would center not on the loom weaving the fabric, but on the fiber supply. This greater traceability of the source of problems could prevent effort being wasted on solving nonexistent problems on the loom.

There is another, less obvious, benefit. If at any stage the quality of the raw fiber improves, then that of the finished fabric should also. A failure to improve would then indicate some problem in the loom. Adjusting the fabric strength for the raw fiber properties makes it possible to recognize improvements that should materialize but don't.

The simplest form of an adjustment is one based on multiple linear regression. For this, suppose that the process measure X_n and covariance vector $\boldsymbol{W}_n$ are connected by a regression model

$$X_n = \boldsymbol{W}'_n\boldsymbol{\beta} + Y_n,$$

where $\boldsymbol{\beta}$ is the vector of regression coefficients connecting $\boldsymbol{W}$ to X and Y_n is the regression residual. Adjusting X_n for $\boldsymbol{W}_n$ is done by using Y_n in place of X_n. Another way to think about this is that X_n is adjusted for $\boldsymbol{W}_n$ by subtracting $\boldsymbol{W}'_n\boldsymbol{\beta}$, the value that X_n "should" have in the absence of the common cause variability Y_n. Usually, we define Y_n to have an in-control true mean of zero. The mean of X_n is incorporated into the "regression" portion of the equation.

We assume that if a problem arises in the process being monitored, it will affect the Y_n term only. Logically, it could not affect $\boldsymbol{W}_n$, since these covariates come from outside the process. As another possibility, it could affect the relationship, for example, by changing the regression relationship between $\boldsymbol{W}$ and X to one with, say, a different slope vector. However, even this sort of change can often be detected by monitoring Y_n.

The strength of a linear relationship like that between $\boldsymbol{W}$ and X is often measured by the squared multiple correlation R^2. This is defined in terms of σ_X^2 and σ_Y^2, the variances of X and of Y, by

$$R^2 = 1 - \frac{\sigma_Y^2}{\sigma_X^2}.$$

In covariate adjustment problems, R^2 measures how much more sensitive control using Y is than control using X without adjustment.

Suppose (mainly to simplify discussion) that $\boldsymbol{W}'_n\boldsymbol{\beta}$ follows a normal distribution, and that its value at one sampling occasion is independent of its value at any other occasion. Suppose some process change occurs and changes the mean of Y_n from 0 to δ. As we assume the regression part of the model does not change, this will change the mean of X by δ also. Our ability to detect this shift is measured by expressing the shift in multiples of the standard deviation. If we make a regression adjustment on X_n and base the control on Y_n, then the standardized shift will be δ/σ_Y. If, however, we ignore the regression correction and just use X_n, then the standardized shift will be δ/σ_X. The ratio of these two standardized shifts is

$$\frac{\delta/\sigma_Y}{\delta/\sigma_X} = (1 - R^2)^{-0.5}.$$

This means that even quite modest multiple correlations can give a valuable magnification in the standardized shift. This in turn means that smaller shifts can be detected quickly.

8.4.1 Example

We illustrate the power of regression adjustment with the ambulatory monitoring data introduced earlier. We mentioned that instrumental issues —

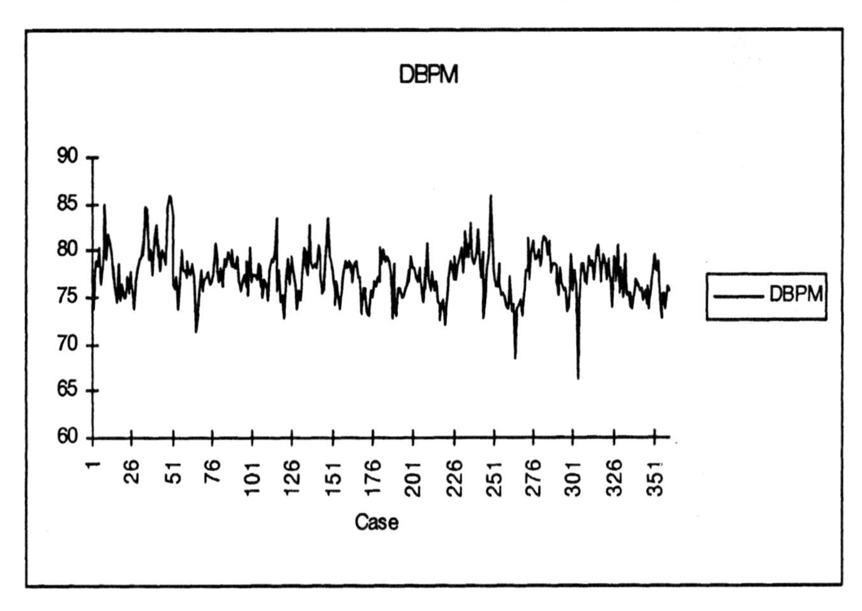

FIGURE 8.4. Time series plot of DBPM.

for example, whether the equipment is calibrated correctly — can be a concern. Regression adjustment can be particularly helpful here. We illustrate this with the early portion of the DBPM series, which is shown in time series form in Figure 8.4. This series looks pretty unremarkable, and in fact had been accepted at face value by the clinicians.

Systolic blood pressure, diastolic blood pressure, and mean arterial pressure are highly correlated. This means that making regression adjustments of any of them for the other two may give substantially better information on the noncommon part of the variability in them.

Figure 8.5 shows the DBPM data following regression adjustment for the SBPM and MAPM MESORS. No formal interpretation is needed; the first 50 or so readings are clearly from a different population than the rest of the series.

On checking back, it was unclear what change in instrumentation or procedures had been made at the end of the first year of data- gathering. The change had not been observed by the clinicians until these analyses were shared with them, and it was difficult to diagnose causes so long after the fact. Doing a multivariate analysis like this, however, clearly revealed an otherwise overlooked change.

8.4.2 SPC use of regression-adjusted variables

The previous example shows a simple but often powerful use of regression-adjusted variables for detecting instrumental problems in one of the process

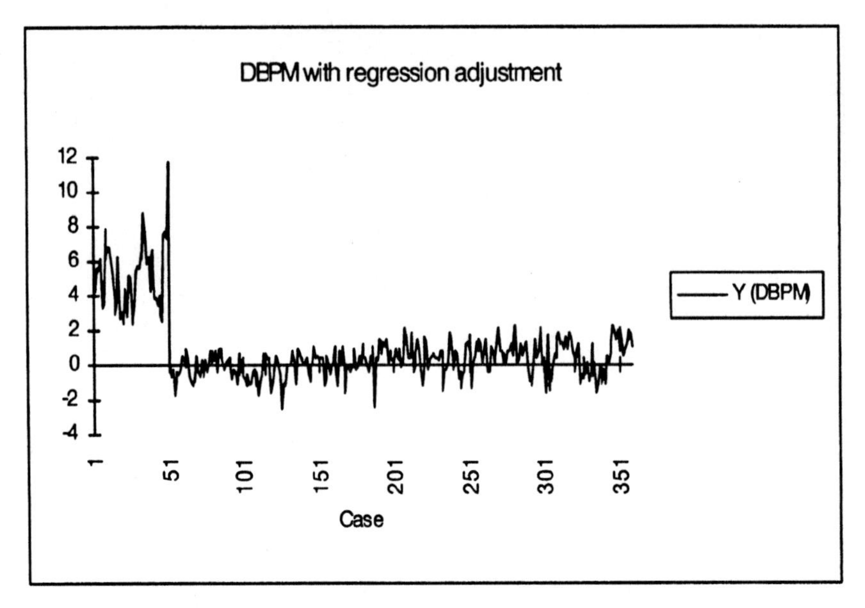

FIGURE 8.5. Time series plot of DBPM with regression adjustment.

measures. Hawkins (1991) gave another similar setting of precious metal assays, where recalibration of one of the measures could mask real process changes.

A number of departures from control can be detected (though not necessarily diagnosed automatically) using the regression-adjusted variables Y_n. For example:

- if there is a step change in mean in which the X_n for a given $\boldsymbol{W}_n$ is larger than it was before the change, then the mean of the Y_n will become positive, and this can be detected by a location CUSUM of the Y_n. A decrease in mean can be detected in the same way.

- If the X_n becomes more erratic in the sense that the variance of X_n for a given $\boldsymbol{W}_n$ increases, this corresponds to an increase in the variance of the Y_n and so can be detected by monitoring the scale of the Y_n. This scale CUSUM will be more effective than the scale CUSUM of the X_n themselves. The amount of improvement is dependent on the ratio of the variances of X and Y - that is, on

$$\sigma_X^2/\sigma_Y^2 = (1 - R^2)^{-1}.$$

- Even a change in the relationship between X_n and W_n may be detectable. If the coefficients $\boldsymbol{\beta}$ change, then the fit of the model using the in-control $\boldsymbol{\beta}$ will deteriorate. Then as the covariates $\boldsymbol{W}_n$ change, the regression adjustment will fail to adjust correctly, and this will tend to make the residuals Y_n more variable. Thus this breakdown in the regression is also likely (although not certain) to leave footprints in the variability of the Y_n.

8.4.3 Example — monitoring a carbide plant

Calcium carbide is used to generate acetylene gas. It is made by passing a strong electric current through a reactor containing a mixture of lime and anthracite (a type of coal). The amount of carbide produced depends on a number of controllable settings in the furnace, but also on the properties of the anthracite and lime. We studied the records of a furnace for approximately 150 shifts, measuring the production X_n in shift n, along with various measures of control settings selected by the operators. We also measured the distribution of particle sizes on the anthracite. Chunks larger than 8mm (the +8 fraction) and those smaller than 4mm (the -4 fraction) seemed to have the most effect on production. We also had records of the limestone screen size, a fairly crude measure of the size of the lime particles being fed in.

The particle sizes of the anthracite and lime are outside the control of the furnace operators, and so they should regard them as covariates.

Using the first 100 shifts to explore the relationship between the production and the feed gave this multiple regression (produced by Statistix 4.1)

```
UNWEIGHTED LEAST SQUARES LINEAR REGRESSION OF PRODUCTION

PREDICTOR
VARIABLES  COEFFICIENT  STD ERROR  STUDENT'S T     P      VIF
CONSTANT     39.2726     10.7156      3.66      0.0004
ANTH8         0.06841     0.02087     3.28      0.0015   1.2
ANTHM4       -0.44659     0.12355    -3.61      0.0005   1.2
LIMESCRN      2.63125     0.99409     2.65      0.0095   1.0

R-SQUARED            0.1876  RESIDUAL MEAN SQUARE (MSE)  121.981
ADJUSTED R-SQUARED   0.1617  STANDARD ERROR OF ESTIMATE  11.0445
```

This regression shows that there is a highly significant, although not very strong, relationship between the carbide production and the covariates +8 anthracite fraction, -4 anthracite fraction, and lime screen size.

Going beyond the calibration period, it was clear that the production had dropped noticeably. The important question is: was this due to some problem in the furnace itself, or might it be because of changes in the anthracite and lime feed? Some graphical checks showed that the anthracite particle size distribution did indeed change over time, and so did the lime screen size.

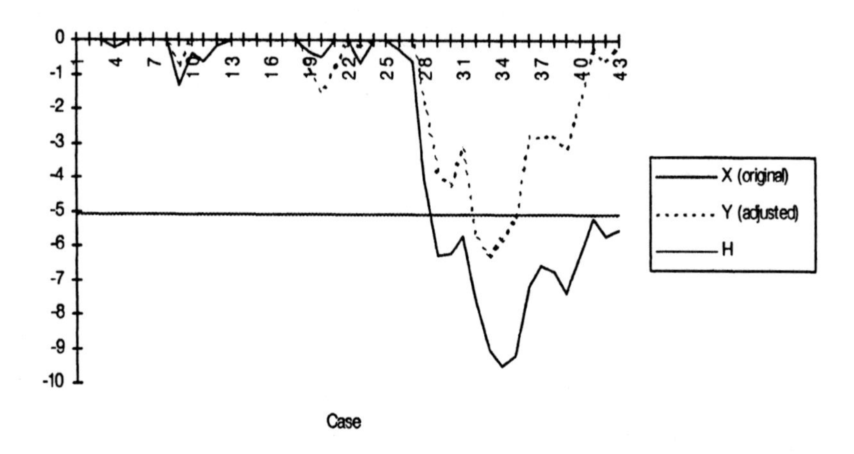

FIGURE 8.6. Original-scale and regression-adjusted production.

This is a natural candidate for covariate adjustment. We used the fitted regression to calculate residuals Y_n for the follow-up period, and made a decision interval CUSUM.

The results are shown in Figure 8.6, which is the downward location CUSUM for the regression-adjusted variables Y_n, along with that of the actual production figures X_n. The reference value k is set to 0.5, and the decision interval of 5.07 gives an in-control ARL of 1,000.

Both the raw production figures and those following regression adjustment cross the decision interval. This means that there does indeed seem to be some special cause within the furnace that led to a production decrease; the decrease could not be attributed to just the changes in the feed material. The CUSUMs also show, however, that the signal given on X_n is much stronger than that from Y_n. This suggests that a substantial part of the downward shift might be due to the raw material.

Following the signal, the regression-adjusted CUSUM quickly returns to within the decision interval, suggesting that the problem within the furnace was short-lived.

8.5 Choice of regression adjustment

These examples have illustrated how regression adjustment can achieve two desirable goals: making the control more sensitive to real shifts by filtering out some components of the variability in the process measurement, and giving more accurate signals by correcting a process measurement for shifts induced by other causative factors outside the process being monitored.

Returning to the multivariate situation, we have a measurement vector $\boldsymbol{X}_n$ of p components, rather than a scalar X_n. The principle of regression adjustment can still be a key tool.

- *Sequential processing problems.* Consider an assembly line operation in which each unit goes through a number of sequential processing steps, each of which is followed by a quality measurement made on the item. In many such processes, the quality as measured at each process step is affected by some or all of those at previous steps. When monitoring each measure, therefore, it is important to distinguish that portion of the process measure that is due to the most recent processing step from that due to the previous steps.

  One obvious way to do this is to make a regression adjustment of *each* process measure for *all* process measures *"upstream"* of it. Thus, for example, the third process measure X_{n3} would be adjusted by regression for X_{n1} and X_{n2}. In this way, a shift seen in X_{n3} can be diagnosed as being due to something in step 3, or as an effect of causes operating upstream. This is a generalization of the carbide production example.

- *Isolated instrument malfunction problems.* The ambulatory monitoring example illustrated another common scenario in multichannel measurement. The different measures made are inherently intercorrelated. It is possible though for one of the measuring channels to malfunction so that its readings are displaced from where they would otherwise have been.

  If the special cause being monitored is of this type — something that will affect a single measurement — then the optimal control is a regression adjustment in which the measurement is adjusted for all the other measurements.

- *Latent variable problems.* In some settings, the process measurements all attempt to capture some small number of elusive underlying ("latent") qualities. In manufacturing color film, for example, there are very many instrumental or sensory measurements that can be made on a sample of film, but these all bear on a smaller number of fitness for use attributes that cannot be measured directly.

  In problems like this, the most effective controls are likely to be based on attempts to estimate the latent variables using the process measurement vectors, and then control using the latent variables. A standard approach for doing this is the use of principal component analysis. Principal components have indeed proved to be very effective for quality control in photographic film manufacture.

8.6 The use of several regression-adjusted variables

Once we get away from the framework of a single process measurement with adjustment for some outside covariates, then we come back to the situation of maintaining a collection of univariate charts, one for each of the regression-adjusted or raw measures being monitored. This situation differs from the MCUSUM of Woodall and Ncube only in the nature of the variables being charted; these will not necessarily be the original process measurements, but may include regression-adjusted variants. This brings up the specter of a large number of charts that need to be maintained and monitored, and also that of a false alarm rate spiraling out of control because each new variable being charted becomes a new source of potential false alarms.

The first of these problems is most effectively handled by "reporting by exception." The multiple CUSUM charts need to be maintained in a computer, with actual charts only being presented for study when they have signaled.

The second problem, that of the spiraling false alarm rate, is controlled by suitable choice of the decision interval h. Assume for simplicity that all charts use the same k and h values. The reference value k is selected on the basis of the size of shift to which you want to tune the CUSUMs, but h is a consequence of decisions about the tolerable in-control ARL.

The run length until the first signal defines the run length of the collection of p CUSUMs regarded as a group. In symbols,

$$RL_{group} = \min_{1 \le i \le p} RL_i$$

Although we can control the individual ARLs of the CUSUMs by suitable choice of h, it is not easy to use this control to fix ARL_{group} exactly. The reason for this is that, in general, there is no restriction on the correlation between the different measures monitored in the CUSUMs.

A heuristic that will generally provide good control, however, is one based on Bonferroni reasoning using the approximation

$$ARL_{group} \approx ARL_{individual}/p.$$

This heuristic then suggests that to make the in-control ARL of the group 1,000, say, one would choose the h so that the individual CUSUMs had in-control ARL of $1000p$.

8.6.1 Example

We illustrate the use of multiple CUSUMs with the ambulatory monitoring data. To check for possible errors such as that affecting the first year's data, we could use a regression adjustment of correcting each of the variables for

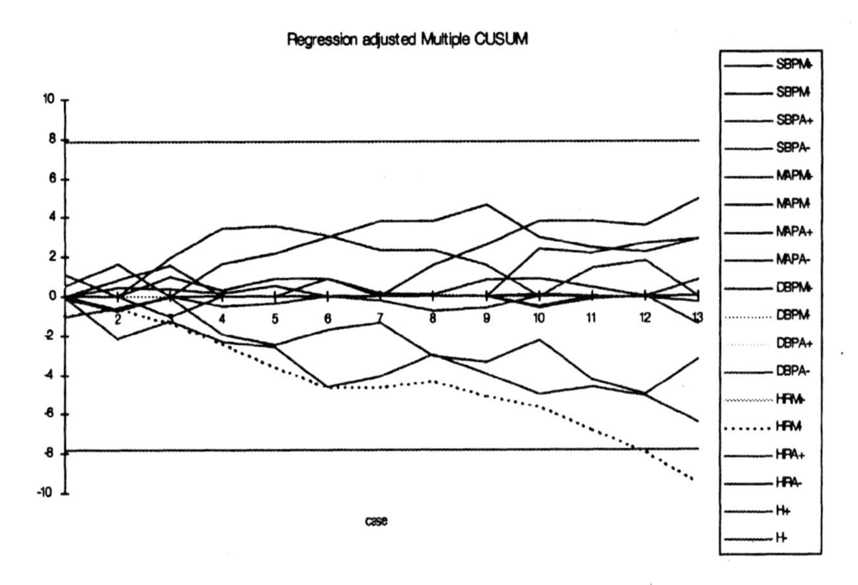

FIGURE 8.7. CUSUMs of all eight regression adjusted variables from the ambulatory monitoring data.

all others; this is the approach particularly indicated if we think one variable may change in mean while the other variables and the regression remain the same.

The CUSUMs of all 8 regression-adjusted variables are shown in Figure 8.7. As we are checking 8 CUSUMs for either upward or downward movement, there are a total of 16 decision intervals being tested at each time, so to hold the overall scheme's in-control ARL in the vicinity of 1,000, we set the decision interval for an in-control ARL of 16,000. With the reference value of $k = 0.5$ standard deviations, ANYGETH reports the needed h as 7.83. The regression-adjusted CUSUM of HRM broke through this decision interval decisively at 13 weeks; at this stage it had been going downward steadily almost from the start of the CUSUM. None of the other measures was anywhere near the decision interval at this time, so the diagnosis of a decreased mean in heart rate MESOR seems clear.

8.7 The multivariate exponentially weighted moving average

There is another way of accumulating information that is particularly useful in multivariate settings: the multivariate exponentially weighted moving average (MEWMA), which is discussed in detail in Lowry et al. (1992).

This is based on the recursions:

$$\boldsymbol{S}_0 = \boldsymbol{\mu}_0$$

$$\boldsymbol{S}_n = \lambda \boldsymbol{X}_n + (1 - \lambda)\boldsymbol{S}_{n-1}$$

and signals if $C_n = (\boldsymbol{S}_n - \boldsymbol{\mu}_0)'\boldsymbol{\Sigma}^{-1}(\boldsymbol{S}_n - \boldsymbol{\mu}_0) > h$.

All that we need to chart is the scalar C_n. If this goes above the control limit h, then a change is signaled, and at that point you go to the full vector $\boldsymbol{S}_n$ to figure out the cause.

The vector $\boldsymbol{S}_n$ is a smoothed estimate of the current mean vector at any time. The constant λ is a smoothing constant lying between 0 and 1. Values of λ near zero give a lot of smoothing and are appropriate for detecting small shifts; larger values of λ discount old information faster, and so are appropriate for rapid detection of large shifts. The control limit h is chosen to set the scheme's in-control ARL at some target value.

The MEWMA has a univariate specialization, the exponentially weighted moving average EWMA, sometimes called the geometric moving average. We have not discussed the univariate EWMA before as it is outside the scope of this book. The EWMA is, however, a method that can be nearly as fast as the CUSUM in detecting step changes, and with the attraction that its value at any time gives an immediate estimate of the current process mean, something that the CUSUM provides only after you measure the slope of the latest segment and add the reference value. The drawbacks of the EWMA are that it is not as fast as the CUSUM at detecting step shifts, and it is not as good for estimating when the step change occurred. As we defined it, the MEWMA is just a collection of EWMAs, one for each of the variables being monitored.

The signal from a MEWMA comes from a T^2-type measure, and so has the same need as does Hotelling's T^2 chart and Crosier's approach for a follow-up to see what the root cause of the signal is. As we saw with the DBPM, it may not be enough to look at the individual components of $\boldsymbol{S}_n$, since the multivariate shift could occur without visibly affecting the individual variables, and so some follow-up involving regression adjustment may be needed.

Figure 8.8 shows the C_n of the MEWMA of the ambulatory monitoring data, using a smoothing constant $\lambda = 0.1$. The chart seems to be moving up steadily, indicating that the process shifted about or very soon after the start of the monitoring period.

8.8 Summary

Multivariate control creates challenging problems of detection and diagnosis of persistent shifts. If the relationship between the variables changes, there is a multivariate shift, even if the individual variables' means do not

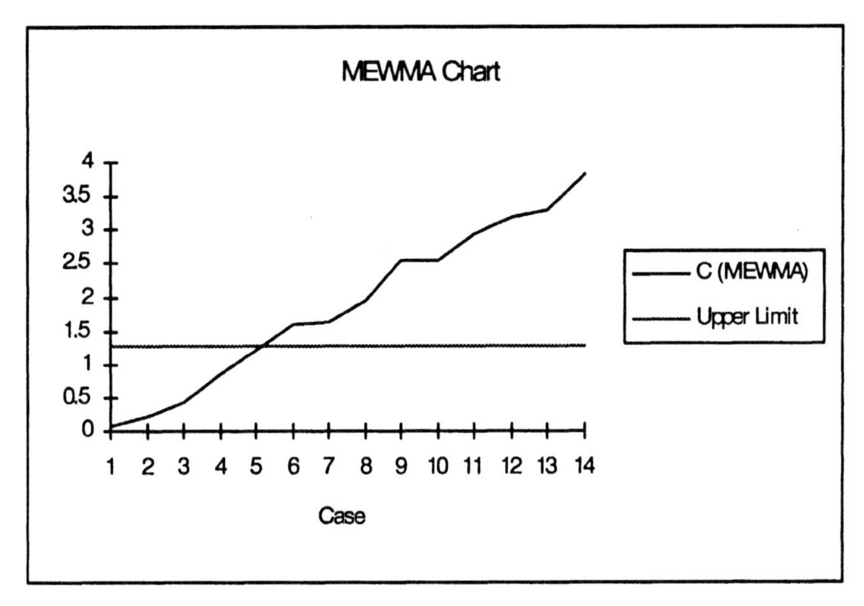

FIGURE 8.8. MEWMA of the ambulatory data.

change; changes of this type defy detection using the original-scale individual measures.

If you have a good idea of how the process is likely to go out of control — that is, of the direction in which the mean vector is likely to move — then Healy's approach of the univariate CUSUM tuned for that direction is an excellent choice. If there are several possible directions, then a collection of univariate CUSUMs tuned for these different directions works well. Regression adjustment gives a general class of summands that make use of process information to look for shifts in likely directions, and in some settings CUSUMs of the principal components are powerful and interpretable.

For problems where you have no good idea of the likely direction of shifts, affine invariant approaches are advisable, and Crosier's proposal, although it has no known optimality properties, seems to work well. The multivariate exponentially weighted moving average is another attractive approach to this problem.

In all multivariate problems, diagnosing the shift can be difficult. A single special cause may lead to a shift in none, one, or more than one of the variables. Regression adjustment can be helpful in resolving these problems.

8.9 Further reading

The Shewhart chart of T^2 is perhaps less familiar than the other Shewhart charts that complement the CUSUM techniques we have been discussing.

Other useful discussion includes Alt (1985). One of the problems of T^2 and of all multivariate methods using the Hotelling metric is interpretation following a signal. Murphy (1987) and Doganaksoy et al. (1991) consider this problem and propose assessing the individual components by univariate t tests and using either a ranking scheme or Bonferroni bounds to choose the component(s) most likely to have gone out of control.

Alwan (1986) develops the multivariate CUSUM for a mean shift in a given direction for a multivariate normal distribution, with the covariance matrix known and fixed. This forms the basis for a CUSUM scheme, which Alwan calls the "MCUSUM" and that parallels Healy's approach.

We mentioned that the T^2 statistic can be triggered by a change in the covariance matrix. This is discussed more fully in Healy (1987).

Principal component analysis is a widely used multivariate methodology with SPC applications. Jackson (1991) provides extensive coverage of the use of principal component analysis of multivariate data, including a sketch of some process control applications although without a discussion of CUSUMs. CUSUM control of the components however is an immediate extension of the Shewhart controls.

Pignatiello and Runger (1990) described a multivariate CUSUM similar in flavor to Crosier's. Where Crosier's shrinks the vector $\boldsymbol{S}_n$ smoothly toward zero, theirs restarts $\boldsymbol{S}_n$ from zero if its norm is sufficiently small. The two proposals turn out to have quite similar performance. They also confirm the inferiority of the CUSUM of T^2. Their proposal to CUSUM transformed variables turns out to be a special case of a CUSUM of principal components.

In our discussion, we have assumed known parameters. Even more than in the scalar case, this is usually no more than a rough approximation to reality. Tracy et al. (1992) replace the χ^2 distribution of T^2 that we have used with properly studentized versions. Unfortunately, although these variants are very helpful for Shewhart charting, they do not extend to multivariate data very easily.

As a somewhat unusual side-benefit of the multivariate approach, Crosier (1986) specialized his multivariate proposal to a scalar CUSUM for the two-sided alternative, and showed a small improvement in performance over using a pair of one-sided DI CUSUMs.

9
Special topics

All progress is precarious, and the solution of one problem
brings us face to face with another problem.
Martin Luther King, Jr.

We conclude this monograph by discussing three special topics.

9.1 Robust CUSUMs

For the most part, our discussion has assumed that the actual data followed the assumed in-control distribution except for the possible step changes in the parameters for which the CUSUM was designed. This is not always the case. To sketch some of the concerns, we talk about a location CUSUM of supposedly normal $N(\mu, \sigma^2)$ individuals data X_n; the major considerations of that problem spill over into other CUSUM applications as well.

The normal distribution can fail in several important ways:

- the actual distribution may be heavier-tailed than normal, so that occasional far-out values occur;
- on occasion the true value may be corrupted and a far-out value occur as a result of some unnoticed procedural error; or
- there may be an isolated special cause that affects an individual reading.

There is no sharp dividing line between these three situations. If procedural errors occur in some stable way, then they become part of the natural distribution of the data, which will then be heavy-tailed. This is quite common in micro-assaying, where mistakes in pipetting out the tiny volumes of fluid used lead to a steady small proportion of substantial outliers. The second and third situations also differ only in how you react to the cause of the error. Pipetting errors are a known source of assaying outliers and do not call for special follow-up unless they start to become more frequent. But an outlier caused by an intermittent failure in an instrument would call for a follow-up: this sort of special cause can be an early warning of a general instrument failure.

An outlying X_n, whatever may be its cause, can cause a signal on the CUSUM; a single X_n greater than $(h^+ - C_{n-1}^+ + k^+)$ will give an out-of-control signal on the upward CUSUM. On the face of it, this is a plus for the CUSUM – that it will detect isolated as well as persistent special causes. But in fact it is not a plus but a drawback. The correct chart for detecting isolated special causes is a Shewhart chart. If the CUSUM signals, it is natural to start looking for a *persistent* special cause. High specificity is a good property in any chart; it is better to detect isolated special causes with the Shewhart chart, and somehow ensure that they do not cause signals on the CUSUM chart.

At the same time, if there is a persistent large shift, the first symptom will be the sudden appearance of a stream of apparent outliers. This means that although the CUSUM should be modified so that it is not triggered by individual outlying observations, it should respond if the outliers are not isolated.

This thinking has led to two ways of "robustifying" the CUSUM. The "robust CUSUM" proposal of Lucas and Crosier (1982) uses an outlier threshold T. If a single observation is more than T standard deviations from μ, it is ignored completely. But if *two successive* observations are more than T standard deviations away and on the same side of μ, then the CUSUM is triggered immediately.

The other approach for making the CUSUM robust to isolated outliers is that of "winsorization." In this approach, outlying observations are "edited" to more central values and these values are then used in the CUSUM.

In the following discussion, we work with the standardized variables

$$U_n = \frac{X_n - \mu}{\sigma}.$$

You choose a winsorizing constant W. If U_n is less than W, then the CUSUM is updated by

$$S_n^+ = \max(0, S_{n-1}^+ + U_n - k).$$

But if W_n is more than W, then the CUSUM is updated by

$$S_n^+ = \max(0, C_{n-1}^+ + W - k).$$

Thus no matter how far X_n may be from μ, the upward CUSUM will go up by no more than $W - k$. The downward CUSUM is defined similarly, and limits the contribution of any case to no more than $-W + k$.

Strictly speaking, since we usually winsorize on both the right and the left, we should modify the discussion of the upward CUSUM by noting that if $U_n < -W$, the upward CUSUM will be updated by

$$S_n^+ = \max(0, S_{n-1}^+ - W - k)$$

but this effect winsorization has of possibly limiting the return of the CUSUM to zero is secondary and from here on we ignore it.

Winsorizing has a lot of intuitive appeal. Provided W is not too large, it will provide protection against outliers, no matter how severe these may be. And provided W is not too small, most observations will be unedited, and so will contribute their full weight of evidence to the CUSUM. Finally, it does not completely ignore any observation, since all observations contribute to the CUSUM.

If the CUSUM is to be winsorized, then its h parameter must be adjusted to keep the in-control run length the same. The GETARL and GETH codes are able to handle the necessary calculations.

The out-of-control ARL of a winsorized CUSUM can be found using in-control calculations. We just look at the case where the mean shifts by δ standard deviations. Then the CUSUM update is:

$$\begin{aligned} S_n^+ &= \max(0, S_{n-1}^+ + \min(U_n, W) - k) \\ &= \max\{0, S_{n-1}^+ + \min(U_n - \delta, W - \delta) - (k - \delta)\}. \end{aligned}$$

Since $U_n - \delta$ is $N(0,1)$, this is the recursion for a CUSUM of standard normal quantities with reference value $k-\delta$ and winsorizing constant $W-\delta$. The decision interval h^+ is unchanged. Our software programs do these calculations as a matter of course.

Putting these together, Table 9.1 shows the necessary h values to get an ARL of 1,000 with normal data and a reference value of $k = 0.5$. Also shown is the run length that results if the mean shifts by $\delta = 1$ standard deviation.

Large values of W correspond to very little editing of data, and so give essentially the same h value that we saw earlier for nonrobust CUSUMs. Winsorizing obviously reduces the sensitivity of the CUSUM for detecting genuine shifts in mean, but as the table shows there is very little loss of performance when winsorizing using values of W of 2 or above.

The other side of winsorizing is the protection it gives against outliers. To give some feeling for this, we calculated the in-control ARL if the mea-

W	h	ARL following shift
1	2.245	13.54
1.5	3.715	11.60
2	4.539	10.91
2.5	4.899	10.63
3	5.023	10.55
3.5	5.060	10.53

TABLE 9.1. h and out-of-control ARL for different W, for a CUSUM of $N(0,1)$ data with in-control ARL = 1,000 and $k = 0.5$.

surements follow the Huber distribution

$$f(x) \propto \begin{cases} \exp(-0.5x^2) & |x| \leq 2 \\ \exp(-|x|) & |x| > 2. \end{cases}$$

This density is normal in the middle, but about 10% of its values fall into the tails, which decay much more slowly than do the tails of the normal. If we do not winsorize, the ARL for this distribution using $k = 0.5, h = 5.072$ is 135, far below the 1,000 that would hold if the data were normal. The in-control ARL using winsorization and the same k, h values is

W	1	1.5	2	2.5	3	3.5
ARL	732	619	472	330	246	199

Even $W = 3$, which would trim just one normal observation in 740, gives a considerable improvement, and the ARL improves steadily as W is decreased. As usual, there is some tradeoff between the protection you get against heavy tails and the price paid in performance if the tails happen to be normal. For these two scenarios (normal and Huber data), values of W around 2 seem to be a good choice as they provide good control of the ARL when the data do have heavy tails, and good sensitivity to genuine shifts when they do not.

Once we decide to winsorize, we need to adjust our choice of h to obtain the desired in-control ARL. Our software programs offer the option to winsorize, and then adjust the ARLs once the winsorization constant has been selected.

Winsorization is a generally useful method, and is not confined to the normal distribution.

9.2 Recursive residuals in regression

SPC methods are often thought of just in their original context of assembly line manufacturing operations. In fact, as some of the examples of the earlier chapters have illustrated, they have uses far beyond these narrow confines. One of the more unusual, perhaps, is as a diagnostic in multiple regression.

Suppose we have a set of multiple regression data $\{Y_i, \boldsymbol{x}_i\}, i = 1, 2, \ldots, n$ that are related by the multiple regression relationship

$$Y_i = \boldsymbol{x}_i'\boldsymbol{\beta} + \epsilon_i,$$

where $\boldsymbol{\beta}$ is the vector of regression coefficients, and ϵ_i is a sequence of true residuals. Under the usual model these residuals are distributed as $N(0, \sigma^2)$ and are independent. Assume that there are p predictors (usually including an intercept) so that both $\boldsymbol{x}_i$ and $\boldsymbol{\beta}$ have p components.

Suppose initially that the data are in time order. Then one of the ways the regression model can fail is that the regression coefficients might change over time, or the residual variance might change over time. If this is a real possibility, we would like to have some diagnostics to know whether it has happened in a particular data set. Recursive residuals are a natural way of addressing this question. They are based on starting out with no data, adding data one case at a time, refitting the regression and using it to predict the next case, and then checking the deviation between the next case's Y and the prediction. This sounds very much like what is done in control charting, where the data are also added one point at a time with a check as each new point is added. It is not surprising then that SPC methods are useful with recursive residuals.

The basic idea of using CUSUMs of recursive residuals is due to Brown et al. (1975). The methodology we outline here uses more advanced CUSUM techniques than were available to those authors in 1975.

9.2.1 *Definition and properties*

Recursive residuals are based on repeatedly fitting the regression model to a steadily growing data set. Some notation is needed. Write $\boldsymbol{Y}_m$ for the m-component column vector made by stacking the Y_i values of the first m cases. Then stack the $\boldsymbol{x}_i$ vectors in sequence as the rows of a design matrix, writing $\boldsymbol{X}_m$ for the $m \times p$ design matrix for the first m cases.

The estimated regression coefficient using the first m cases is then given by

$$\hat{\boldsymbol{\beta}}_m = (\boldsymbol{X}_m'\boldsymbol{X}_m)^{-1}(\boldsymbol{X}_m'\boldsymbol{Y}_m).$$

Using this estimated regression coefficient to predict the following case will give the prediction $\boldsymbol{x}_{m+1}'\hat{\boldsymbol{\beta}}_m$. Under the regression model, the difference between the actual Y_{m+1} and this prediction will follow a normal distribution with mean zero and variance $\sigma^2(1 + \boldsymbol{x}_{m+1}'(\boldsymbol{X}_m'\boldsymbol{X}_m)^{-1}\boldsymbol{x}_{m+1})$.

The $(m+1)$th recursive residual is then defined by

$$r_{m+1} = \frac{Y_{m+1} - \boldsymbol{x}_{m+1}'\hat{\boldsymbol{\beta}}_m}{\sqrt{1 + \boldsymbol{x}_{m+1}'(\boldsymbol{X}_m'\boldsymbol{X}_m)^{-1}\boldsymbol{x}_{m+1}}}$$

for $m = p, p+1, \ldots, n-1$. Under the normal model, these are independently distributed as $N(0, \sigma^2)$. The study of recursive residuals then consists of checking the sequence of r_m for evidence of departure from this normal model. Particularly interesting examples include

- r_m may be an outlier, indicating that case m is incompatible with its predecessors;
- the mean of the r_m may depart from zero. This is a possible outcome from a change in the regression from the earlier portion of the data to a later portion; and
- the spread of the r_m may change. This could be because of a change in variance in the original underlying true residuals ϵ_m but can also be caused by a change in a true regression coefficient. If the true coefficient of some predictor changes somewhere in the sequence, then the subsequent correction made for that predictor will not account for its effect correctly, and so variation in the predictor will lead to increased random variability in the recursive residuals.

All three of these departures can be checked using standard SPC methodologies. A Shewhart chart of the r_m will highlight an individual outlying case, a location CUSUM will show up systematic changes in the mean, and a scale CUSUM will show up systematic changes in the residual variance.

As we are concerned with CUSUMs here, we do not say any more about the Shewhart chart of the recursive residuals, but this should not be taken to mean we think this chart using the recursive residuals is unimportant.

Since the data set of a regression problem is generally of some fixed and often quite small size, it is usually unrealistic to suppose that we have available a large calibration data set to estimate σ^2, and so we should analyze using short-run methods such as self-starting CUSUMs.

There are some subtleties that suggest using a more specialized self-starting CUSUM than that set out in Chapter 7. First, although the standard deviation σ of the true residuals is not known, the in-control mean is zero and therefore is not unknown. Thus we should use a hybrid self-starting approach in which we use the fact that the target in-control mean is known and confine the learning to the standard deviation.

Write $W_m = \sum_{j=p+1}^{m} r_j^2$. The obvious estimator for the running variance after m observations is

$$\hat{\sigma}_m^2 = W_m/(m-p).$$

We also need a "pivotal" whose distribution does not depend on any unknown parameters. An equally obvious choice for the pivotal on case m is

$$T_m = \frac{r_m}{\hat{\sigma}_{m-1}}.$$

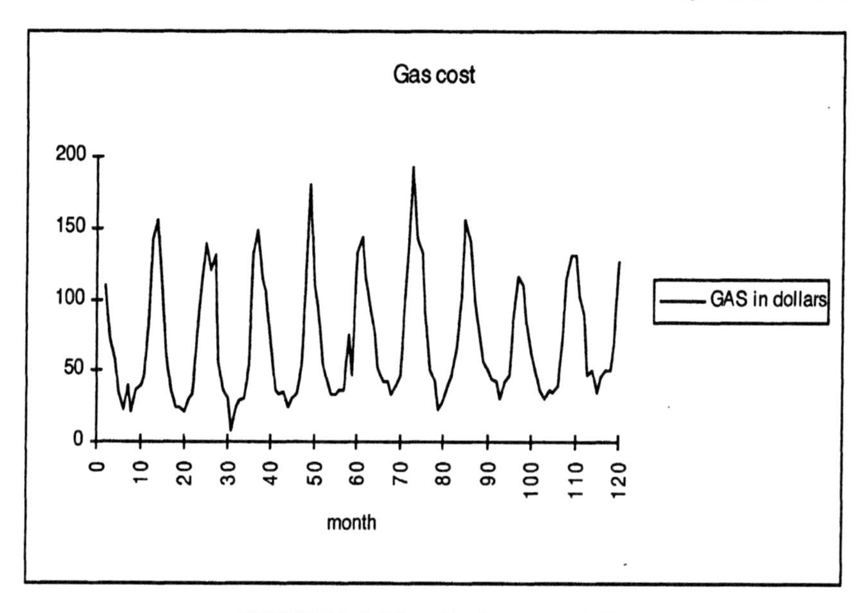

FIGURE 9.1. Monthly home gas bills.

This follows a Student's t distribution with $m-1$ degrees of freedom and the successive terms are independent. The T_m can then be converted to a sequence of standard normals

$$U_m = \Phi^{-1}\left(F_{m-1}(T_m)\right)$$

which can be CUSUMmed to monitor possible shifts in the mean of the recursive residuals.

This deals with control on the mean of the recursive residuals. For control of their variance, we can CUSUM the U_m^2, which follow a χ^2 distribution with one degree of freedom.

The other special feature of recursive residuals is that, since the data set is of a fixed size, there is no logical reason to order the data from earliest to latest; it is equally defensible to order them from latest to earliest. For any data set, this means that you can compute two different sets of recursive residuals: the "forward" recursive residuals given by starting with the first observation and then successively adding the newer observations; and the "backward" recursive residuals given by starting with the last observation and then successively adding the earlier ones. These two sets of recursive residuals give different but complementary pictures of the data.

We illustrate these points with a small case study.

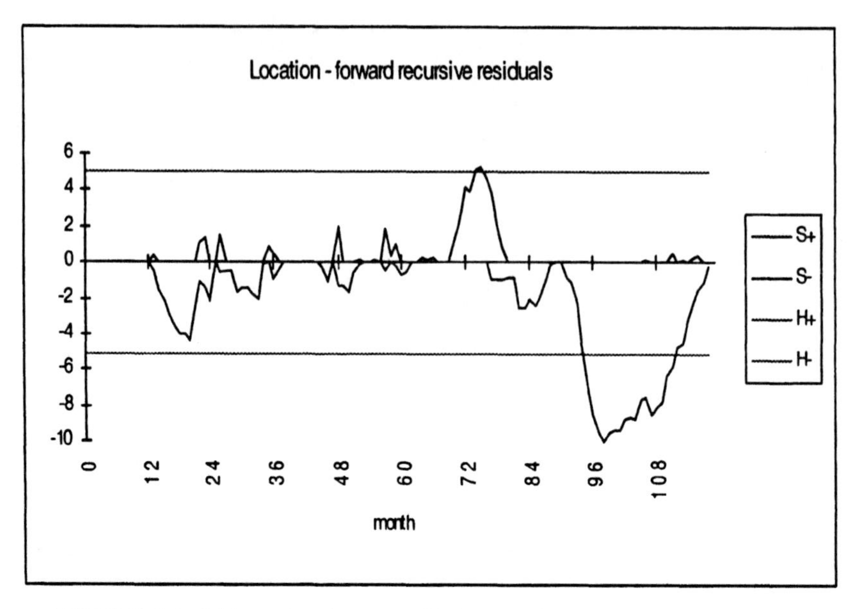

FIGURE 9.2. Self-starting location CUSUM of forward recursive residuals.

9.2.2 *Example*

Figure 9.1 shows the monthly gas bill of a home over a ten-year period. The amounts are clearly strongly seasonal, and it is hard to see whether there is any other structure in them.

There are many ways to analyze a data set such as this. It could be done, for example, using Box–Jenkins ARIMA methods, and this might be particularly appropriate if forecasting were our main concern. We chose to model it using regression instead. The predictors were four harmonics of a Fourier series using the day of year as the basic cycle. Figure 9.2 shows the location CUSUM of the forward recursive residuals. In making it, we used the reference values $k^+ = k^- = 0.5$ for which $h^+ = h^- = 5.072$ would give an in-control ARL of 1,000. The upward CUSUM remained within the decision interval for the whole period. The downward CUSUM broke out of the decision interval decisively at month 96, having left the axis around month 80. The downward move persisted until month 100, after which the CUSUM turned back up toward the axis. The turn up after month 100 does not necessarily mean that the last observations fitted the same model as those prior to month 92; rather it may just be a result of the way the recursive regression adapts to the new observations from the different regime, and the way the self-starting CUSUM adapts to any increase in the residual variance.

The CUSUM of the backward recursive residuals is shown in Figure 9.3. Reading this CUSUM is a bit different than any of the other CUSUMs

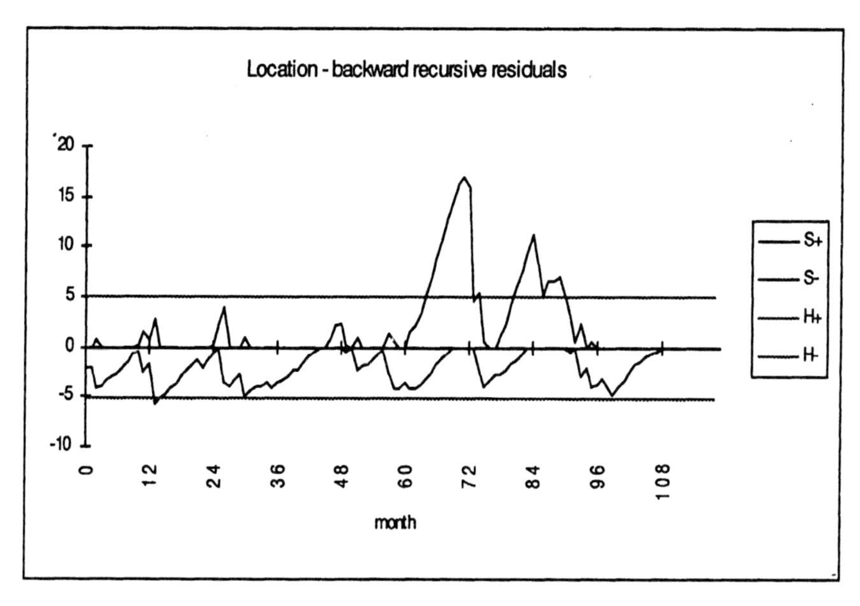

FIGURE 9.3. Self-starting location CUSUM of backward recursive residuals.

we have shown: the accumulation is from right to left rather than left to right. We did this reversal of direction so that the "month" scale could still be read easily, and so that the forward and backward CUSUMs could be compared more easily. One immediately visible difference between the two CUSUMs is that the forward CUSUM does not have any points for the first 10 months, as these months are needed to get the first regression fitted. The backward CUSUM has points for these months (the last to be accumulated into the regression), but not for the last 10 months, as these months were needed to get the first backward regression.

The backward CUSUM shows a strong upward signal starting at month 94 and signaled at month 89. The CUSUM carries on up all the way to month 70 before turning back toward the axis. The upward signal at observation 89 is the mirror image of the downward signal from the forward CUSUM, and confirms that there are at least two regression models — one fitting the earlier data and one the later.

In the face of these substantial location shifts, the scale CUSUMs are not particularly interesting, since the shift in location will generate a shift in scale also. As the scale CUSUMs did not show any features that would indicate a change in variance itself (rather than a change induced by the shift in location) these CUSUMs have been omitted.

9.3 Autocorrelated data

In Chapter 1, we saw how quite modest autocorrelation could have a dramatic impact on the ARL of the CUSUM. Positive autocorrelation leads to a sharp increase in the number of false alarms, whereas negative autocorrelation all but eliminates false alarms but must presumably pay for this in reduced performance. This raises the issue of how to handle data that show measurable autocorrelation.

One reasonable approach can be used when the dependence can be described by a Box–Jenkins autoregressive-moving average (ARMA) model. Under this model, the series X_t can be described by

$$\begin{aligned} X_t &= A\mu + \phi_1 X_{t-1} + \phi_2 X_{t-2} + \ldots + \phi_p X_{t-p} \\ &\quad + e_t + \theta_1 e_{t-1} + \ldots + \theta_q e_{t-q}, \end{aligned} \tag{9.1}$$

where the innovations e_t are "white noise" - independent and identically distributed $N(0, \sigma^2)$ variables. We write ARMA(p, q) as a shorthand for this model. There are p autoregressive terms connecting X_t with its p most recent predecessors, and $q + 1$ moving average terms incorporating the innovation e_t and its q most recent predecessors. The (for our purposes unimportant) constant A is there so that the mean of the X_t will be μ.

If $p = 0$, the series is a pure moving average (MA), and if $q = 0$ it is a pure autoregression (AR). Any ARMA can be re-expressed as a pure MA or as a pure AR by using Equation 9.1 to write

$$\begin{aligned} X_{t-j} &= A\mu + \phi_1 X_{t-j-1} + \ldots + \phi_p X_{t-j-p} \\ &\quad + e_{t-j} + \theta_1 e_{t-j-1} + \ldots + \theta_q e_{t-j-q} \end{aligned}$$

to successively eliminate the X_{t-j} terms on the right of 9.1, or as

$$\begin{aligned} e_{t-j} &= X_{t-j} - A\mu - \phi_1 X_{t-j-1} - \ldots - \phi_p X_{t-j-p} \\ &\quad - \theta_1 e_{t-j-1} - \ldots - \theta_q e_{t-j-q} \end{aligned}$$

to eliminate the e_{t-j} terms. In each case, the term is "eliminated" only in the sense that it is replaced by an older term with a smaller coefficient; if we carry this on indefinitely we will have "infinitely old" terms with infinitesimal coefficients. The resulting pure models will generally have infinite order, but for practical purposes it may be possible to truncate them after a manageable number of terms.

We concentrate for clarity on the pure AR(1), or Markov process. This is perhaps the most common model seen in nonindependent nonseasonal time series, and the broad principles illustrated are the same for any ARMA model. A much fuller discussion is given in Yashchin (1993). The AR(1) model can be written as

$$(X_t - \mu) = \phi(X_{t-1} - \mu) + e_t, \tag{9.2}$$

where we have dropped the subscript on ϕ since there is only one of them. The innovations e_t are independent $N(0, \sigma^2)$ variables. Since we can rewrite X_t as an infinite MA

$$X_t = \mu + \sum_{j=0}^{\infty} \phi^j e_{t-j}$$

the series X_t is itself normal, with mean μ and variance

$$\sigma_X^2 = \sigma^2 \sum_{j=0}^{\infty} \phi^{2j}$$

$$= \frac{\sigma^2}{1 - \phi^2}.$$

The correlation between X_t and X_{t-j} is ϕ^j. This exponential decay in the autocorrelations is the hallmark of a Markov process.

This deals with the in-control modeling of an AR(1). There are many ways in which this process could go out of control. For example:

- the mean of X_t could go from μ_0 to some other constant level μ_1;
- the mean of the innovations e_t could go from zero to some constant non-zero level δ; or
- the variance of the innovations could go from σ_0^2 to some other value σ_1^2.

Note that there are two different step mean shifts that could occur: a step change in X_t or one in e_t. These do not lead to the same outcome. If the innovation mean changes from 0 to δ at some instant, then the mean of X at time m units later will change from μ to

$$\mu + \delta \sum_{j=0}^{m} \phi^j = \mu + \delta \frac{1 - \phi^{m+1}}{1 - \phi}.$$

So the mean of X_t does not have a step change, but moves up toward the asymptote $\mu + \delta/(1 - \phi)$.

Conversely, a step change of Δ in the mean of X_t corresponds to the mean of the first e_t changing from 0 to Δ and the mean of all subsequent e_t changes to $\Delta(1 - \phi)$. If ϕ is close to zero, this is quite like a permanent step change of Δ in the mean of the e_t; at the other end of the correlation scale, if ϕ is close to 1, then the step change in X_t comes about primarily from a shift in a *single* innovation.

Once having estimated ϕ, we can easily convert the X_t series into the series

$$Z_t = X_t - \phi X_{t-1} = (1 - \phi)\mu + e_t$$

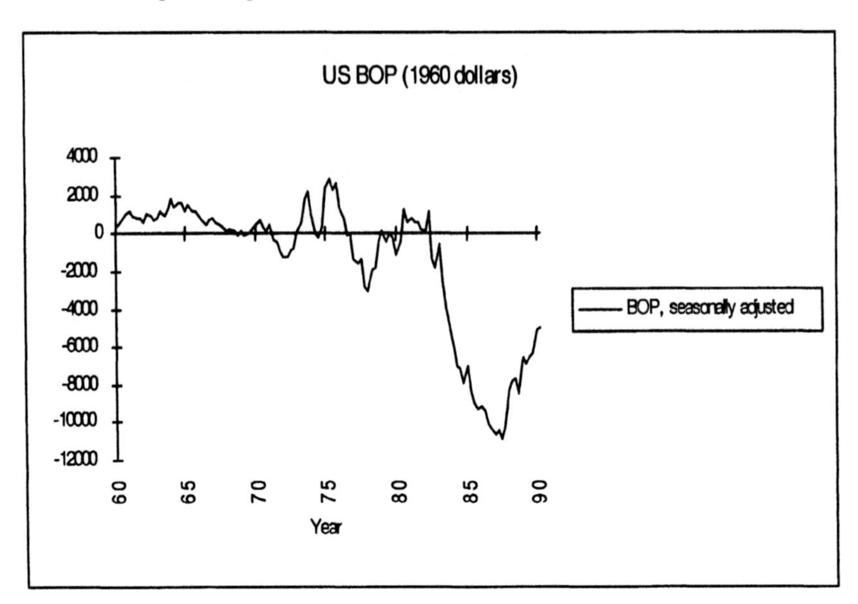

FIGURE 9.4. Quarterly balance of payments.

which, while the process is in control, will be $N\{(1-\phi)\mu, \sigma^2\}$. Applying CUSUM location control to the sequence Z_t will give the optimal diagnostic for a step shift in the mean of the e_t. Applying CUSUM scale control will give the optimal diagnostic for a step shift in the variance of the e_t. CUSUM control on the Z_t will not give the best control for a step change in X_t, but will come quite close to the theoretical best if ϕ is not too large. If ϕ is large, then Shewhart control, looking for individual large Z_t is the indicated approach. Either way SPC of the independent sequence Z_t will provide good control of the original X_t sequence while sidestepping the problem of the serial correlation.

9.3.1 Example

An example may help show how the approach works. The US balance of payments (BOP) has been a source of much concern in recent years. Figure 9.4 — the quarterly BOP from the first quarter of 1960 through the second quarter of 1990 — shows why. The massive deficit of the 1980s is obvious even without statistical analysis. We use this data set anyway to show some features that might be less obvious. In doing so, we want to emphasize that this is not intended to be a definitive econometric analysis, but an illustration of the use of CUSUM technology on an ARMA data set.

Using the first 10 years of the series for calibration gives the autocorrelations

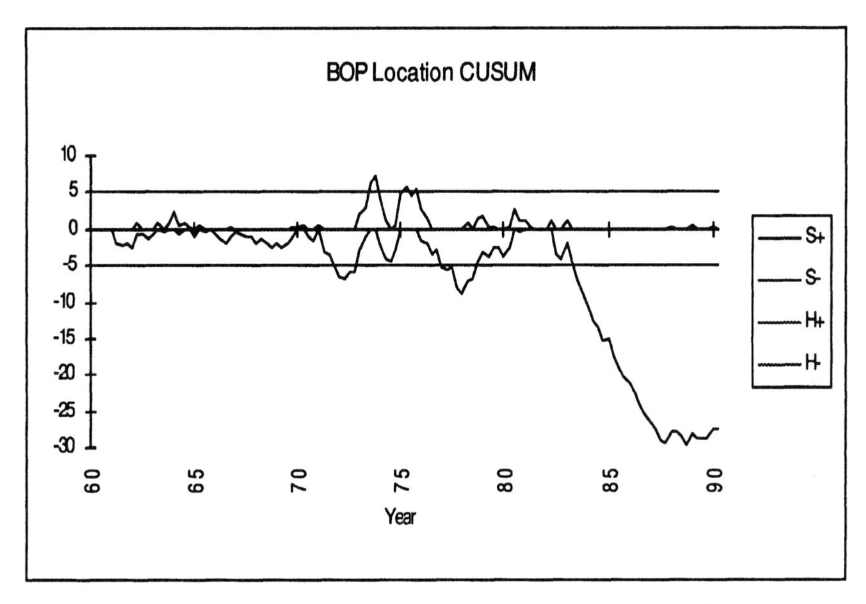

FIGURE 9.5. Location CUSUM — balance of payments.

Lag	1	2	3	4	5
Autocorrelation	0.853	0.734	0.631	0.485	0.410

Under the Markov model with ϕ estimated by the first autocorrelation of 0.853, the subsequent autocorrelations should be 0.727, 0.621, 0.529, and 0.452, respectively. Without doing any formal testing, the actual figures seem to match these quite closely. So we use the AR(1) with $\phi = 0.853$ as a model for the process and calculate the Z_t for the whole series.

If we accept the AR(1) model, there are three parameters to be estimated; ϕ, μ, and σ. Using the assumed value $\phi = 0.853$ still leaves two parameters to be estimated. As the 10-year calibration series used contained only 40 observations, it is hard to be comfortable with plugging in the sample estimates of these parameters and treating them as known, so we study the Z_t using self-starting methods.

The self-starting location CUSUM using $k = 0.5$ is given in Figure 9.5. A decision interval of 5.07 would correspond to an in-control ARL of 1,000, and seems reasonable. The overwhelming visual feature of the CUSUM is the downward shift starting in the second quarter of 1982 and persisting right through to the end of the series. This however is not the only signal — the CUSUM went through the upper decision interval decisively in late 1973 and through the lower decision interval in 1972 and again in 1977.

Leaving the location CUSUM aside for the moment, we also computed scale CUSUMs using the reference values $k^+ = 1.5$ and $k^- = 0.75$. These are intended to monitor doubling and halving of the innovation variance,

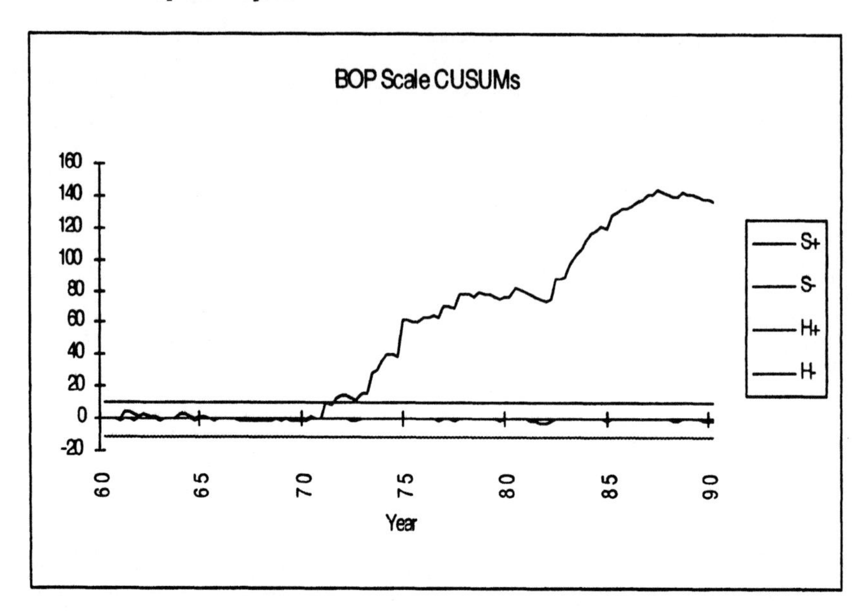

FIGURE 9.6. Scale CUSUMs — balance of payments.

respectively. The scale CUSUMs are shown in Figure 9.6. An upward decision interval of $h^+ = 10.88$ and a downward decision interval of $h^- = 11.05$ would each give an in-control ARL of 1,000. There is no action whatever on the downward scale CUSUM, but the upward CUSUM punched through the 10.88 level in the fourth quarter of 1971, indicating a variance increase starting in the third quarter of 1970. From there on, the scale CUSUM went on rising steadily. Apparently, after a relatively tranquil decade in the 1960s, the balance of payments became much more erratic. As the 1970s were the decade of oil crises, this is not surprising, although it is interesting that the effect of the turmoil is more in the scale of the innovations than their location.

As usual, there is some interference between the two shifts. The strong signal on the scale makes one wonder whether the three weaker location signals in 1972, 1973, and 1977 were side-effects of the variance increase rather than legitimate signals of location shifts. The huge downward location shift in 1982, conversely, may be the cause of some of the continuing upward movement on the scale chart.

We have been concentrating on CUSUMs of the innovations. On the location side, since ϕ is large, a step change in the mean of Z_t would correspond to a drift in the mean of X_t toward a new asymptote. The individual Z_t should also be looked at using Shewhart ideas. Although it is not all that apparent on the CUSUMs, the innovation for the first quarter of 1975 (over 2 billion dollars) yielded a normal equivalent of $U_n = 4.87$, an enormous value surrounded by quite modest values. Since ϕ is quite close to

1, this is the sort of signal that goes with a step change in the mean of X_t. The original sequence does indeed show this higher plateau in the following few quarters, although it is drowned out visually by the later deficits.

As a final comment, we have been analyzing these innovations using self-starting CUSUMs. As these adapt to the changing characteristics of the data, they should really be restarted every time there is a substantial change in the data, otherwise they tend to accommodate to the change and stop signaling it. This general concern does not seem to hold here; these CUSUMs have shown no sign of toning down their signals, but have continued to signal vigorously. Had this not been the case, it would have been a good idea to follow the standard practice of restarting the self-starting scheme after each confirmed shift in mean or variance.

A full econometric analysis (which this is not) would need to take account of more than just the time series itself. Still, this analysis may provide helpful insights and generate useful hypotheses for a fuller study.

9.4 Summary

- Winsorization allows CUSUMs to be robust against certain model departures, namely, outliers, and a misspecified and heavy-tailed distribution. Winsorization is incorporated into our software as an option.
- Self–starting CUSUMs can be used to diagnose regression recursive residuals to detect outliers, mean shifts, and variance shifts.
- CUSUMs can be adapted to deal with autocorrelated data.

9.5 Further reading

9.5.1 Time series

We have chosen to concentrate on the approach of modeling the time series and CUSUM the innovations. If we are worried that the process mean of the data has shifted, it is more natural to CUSUM the readings themselves, and try to take care of the impact of the autocorrelation on the run length. Johnson and Bagshaw (1974) and Bagshaw and Johnson (1975) provided a valuable asymptotic analysis of CUSUMming autocorrelated data. The first paper showed that two important time series models – the autoregression of order 1 (the Markov process) and the moving average of order 1 — and one special case — $k = 0$ could be handled asymptotically by a simple rescaling, dividing the summand by a scaling constant defined by the coefficients of the time series model.

The term "process control" means different things to different people. To statisticians and quality professionals, it generally refers to the type of monitoring by control charts that this book discusses. To many engineers, it refers to the use of on-line monitoring and feedback to continuously adjust control settings to try to keep a process on track. These two methodologies are starting to cross-fertilize: see for example Box and Kramer (1992), Vanderwiel et al. (1992), and Montgomery et al. (1994).

Further discussion of control of autocorrelated data is given by Montgomery and Mastrangelo (1991) and discussants.

9.5.2 Score methods

Along the way, we have discussed a number of specific applications where we transformed our data prior to summation. There are a number of proposals to accumulate various "scores" in order to avoid distributional or other problems with the data. Munford (1980), for example, proposed transforming the process measure X_n to a "score" Y_n, set to 0, 1 or -1 according to whether X_n lay between k_1 and k_2, was larger than k_2, or was smaller than k_1, and CUSUMming the score Y_n. This method can be used for control on any quantity whose in-control statistical distribution is known, as is needed to calculate the probability that Y_n takes on each of its three possible values. The ARL of the CUSUM depends on the distribution of X_n only through these three probabilities, making this CUSUM robust to any imperfections in the assumed distribution that do not affect these probabilities. McGilchrist and Woodyer (1975) use a scoring of 1 and -1 for data, respectively, above and below the median to provide a truly non-parametric CUSUM. Ncube and Woodall (1984) refined Munford's cumulative score scheme by adding a large score for values that would lie outside Shewhart control limits, giving a scheme with combined Shewhart–CUSUM performance and improving response to large shifts. Xiao (1992) went beyond Ncube and Woodall's suggestion, using as scores the midpoints of the classes of a histogram centered at μ and with bin width σ. This scheme (which really consists of using rounded process readings) is nearly as effective for normal data as CUSUMming the actual readings (that is, the rounding has little impact on performance), but of course loses the simplicity of the Munford scheme.

9.5.3 Robustification

A proposal by Rocke (1989) is suitable for robust CUSUMming when you have rational groups, rather than individual observations. His suggestion is to replace the sample mean by a "trimmed" mean. This has the desired end result of protecting the CUSUM from a small proportion of outliers.

A fuller comparison of winsorization and the 'ignore one signal on two' approach to robust CUSUMs is given in Hawkins (1993a).

9.5.4 Recursive residuals

Galpin and Hawkins (1984) and Hawkins (1991) discuss a number of extensions to the Brown et al. (1975) proposals.

10
Software

Software for CUSUM design calculations is available from many sources. Yashchin (1986) describes an APL2 package with the capability of solving many design and ARL problems. At the time of this writing, this software is available on request from the author (yashchi@watson.ibm.com) and is scheduled to be put onto a Web site.

The *Journal of Quality Technology* has published a number of CUSUM programs. These can be accessed through (www.stat.cmu.edu), the **statlib** repository , along with the smaller number of CUSUM programs that have appeared in *Applied Statistics.*

In addition to these sites, the CUSUM Web site of the School of Statistics, University of Minnesota (www.stat.umn.edu) has been designed as a companion to this book and contains all the data sets used in the book, along with software to find ARLs and decision intervals. The software uses Hawkins (1992) CUSARL general-purpose ARL code, with front-end code implementing different data distributions.

10.1 Programs and templates

The associated Web site provides two main programs and several Microsoft Excel templates. The following description is at the time of writing; we expect to add new procedures and new capability to the existing procedures in the future.

Please read the disclaimer about the software in the Preface. We make no warranties about this software.

ANYARL.EXE is a master program that combines most of the ARL routines used in the examples in the text. It can find the ARL for a CUSUM scheme where the reference value and decision interval are known. It allows winsorizing and fast initial response (FIR) CUSUMs.

ANYGETH.EXE is a second program that finds the decision interval (h) for many CUSUM schemes. It requires the user to choose the distribution and to specify the reference value k. It accommodates winsorization and fast initial response (FIR) CUSUMs.

There are five Excel templates provided for normal CUSUMs. They allow CUSUM schemes for normal mean and variance under both the known parameter and self-starting CUSUM procedures. They are:

- DOCUSUM.XLS, a generic CUSUM plotter,
- NORMAL.XLS, for normal means and variances,
- SSNORMAL.XLS, for self starting charts for normal means and variances,
- POISSON.XLS, for Poisson data,
- SSPOISSON.XLS, for self starting charts for Poisson data,
- additional templates as we develop them.

These spreadsheets include Visual Basic code to automate the production of the charts, and include dialogs to elicit the parameters. The spreadsheets are copywritten by D. M. Hawkins and D. H. Olwell, but permission is granted for their use by purchasers of this monograph. Other usage requires permission in writing from the authors.

10.2 Data files

We provide the data files used in this book in ASCII and Excel format. The data files are named by the table in which they were displayed: *TBL9-1.xls* and *TBL9-1.txt*, for example. Files for figures are provided in a similar format: *FIG9-1.xls*.

Additional data files, if any, will be catalogued at the website under the file *moredata.htm*.

11
References

[1] Alwan, Layth C. (1986) CUSUM Quality Control – Multivariate Approach. *Communications in Statistics B.* Vol. 15. No. 12. 3531-3543.

[2] Alt, Frank B. (1985) Multivariate Quality Control. *Encyclopedia of the Statistical Sciences.* Ed N. L. Johnson, S. Kotz and C. R. Read. Vol. 6. 111-122.

[3] Alt, Frank. B., Heikes,R. G., and Montgomery, Douglas C. (1986)Economic Design of Chi-Square Control Chart for Multivariate Processes. Presented paper, Annual meetings of the American Statistical Association. Chicago.

[4] Assaf, David, and Ritov, Ya'acov. (1989) Dynamic sampling procedures for detecting a change in the drift of Brownian motion: A non-Bayesian model. *The Annals of Statistics.* Vol. 17. 793-800.

[5] Bagshaw, Michael, and Richard A. Johnson. (1975) The effect of serial correlation on the performance of CUSUM tests II. *Technometrics.* Vol. 17. 73-80.

[6] Bagshaw, Michael, and Johnson, Richard A. (1975) The influence of reference values and estimated variance on the ARL of CUSUM tests. *Journal of the Royal Statistical Society, Series B, Methodological.* Vol. 37. 413-420.

[7] Banzal, R. K. and Papantoni-Kazakos, P. (1986) An algorithm for detecting a change in a stochastic process. *IEEE Transactions on Information Theory.* Vol. IT-32. 227–235.

[8] Barnard, George A. (1959) Control Charts and Stochastic Processes. *Journal of the Royal Statistical Society, Series B.* Vol. 21. No. 2. 239-257.

[9] Basseville, M. and Nikiforov, I. (1993) Detection of Abrupt Changes. Englewood Cliffs, NJ: Prentice Hall.

[10] Bather, John A. (1963) Control Charts and Minimization of Costs. *Journal of the Royal Statistical Society, (B).* Vol. 25.

[11] Bauer, Peter, and Hackl, Peter. (1978) The use of MOSUMs for quality control. *Technometrics.* Vol. 20. 431–436.

[12] Bauer, Peter, and Hackl, Peter. (1980) An extension of the MOSUM technique for quality control. *Technometrics.* Vol. 22. 1-7.

[13] Baxter, L. A., Coutts, S. M., and Ross, G. A. (1980) Applications of linear models in motor insurance. *Proceedings of the 21st International Congress of Actuaries.* Zurich. 11-29.

[14] Bissell, Alfred F. (1969) CUSUM techniques for quality control (with discussion). *Applied Statistics.* Vol. 18. 1-30.

[15] Bissell, A. F. (1973) Process monitoring with variable element sizes. *Applied Statistics.* Vol. 22. 226-238.

[16] Bissell, Alfred F. (1973) Process monitoring with variable element sizes. *Applied Statistics.* Vol. 22. 226-238.

[17] Bissell, Alfred F. (1979) A semi-parabolic mask for CUSUM charts (Corr: V30 p77). *The Statistician.* Vol. 28. 1-8.

[18] Bissell, A. F. (1984) The performance of control charts and cusums under linear trend. *Applied Statistics.* Vol. 33. 145-151.

[19] Bissell, Alfred F. (1984) The performance of control charts and cusums under linear trend. *Applied Statistics.* Vol. 33. 145-151.

[20] Bissell, Alfred F. (1984) Estimation of linear trend from a CUSUM chart or tabulation. *Applied Statistics.* Vol. 33. 152-157.

[21] Bourke, Patrick D. (1991) Detecting a shift in fraction nonconforming using run-length control charts with 100 percent inspection. *Journal of Quality Technology.* Vol. 23. 225-238.

[22] Bourke, Patrick D. (1992) Performance of cumulative sum schemes for monitoring low count-level processes. *Metrika.* Vol. 39. 365-384.

[23] Box, George E. P. and Jenkins, G. M. (1976) *Time Series Analysis: Forecasting and Control* 2nd ed. San Francisco: Holden-Day.

[24] Box, George E. P. and Kramer, T. (1992) Statistical Process Monitoring and Feedback Adjustment – A discussion. *Technometrics.* Vol. 34. 51 –267.

[25] Box, George E. P., and Ramirirez, J. (1992) Cumulative score charts. *Quality and Reliability Engineering International.* Vol. 8. 17-27.

[26] Brook D., and Evans, D. A. (1972) An approach to the probability distribution of CUSUM run length. *Biometrika* Vol. 59. 539–549.

[27] Brown, R. L., Durbin, J. and Evans, J. M. (1975) Techniques for testing the constancy of regression relationships over time. *Journal of the Royal Statistical Society, Series B, Methodological.* Vol. 37. 149-192.

[28] Champ, Charles W. and Woodall, William H. (1987) Exact results for Shewhart control charts with supplementary runs rules. *Technometrics.* Vol. 29. 393-399.

[29] Chhikara, R. S. and Folks, J. L. (1977) The Inverse Gaussian Distribution as a Lifetime Model. *Technometrics* Vol. 19. No. 4. 461-468.

[30] Chhikara, R. S. and Folks, J. L. (1989) *The Inverse Gaussian Distribution.* New York: Marcel Dekker.

[31] Chiu, W. K. (1974) The economic design of CUSUM charts for controlling normal means. *Applied Statistics.* Vol. 23. 420-433.

[32] Chiu, W. K. (1974) The economic design of CUSUM charts for controlling normal means. *Applied Statistics.* Vol. 23. 420-433.

[33] Crosier, Ronald B. (1986) A new two-sided cumulative sum quality control scheme (Corr: V31 p503). *Technometrics.* Vol. 28. 187-194.

[34] Crosier, Ronald B. (1988) Multivariate Generalizations of Cumulative Sum Quality-Control Schemes. *Technometrics* Vol. 30. No. 3. 291-303.

[35] Dawid, A. P. (1979) Conditional Independence in Statistical Theory. *Journal of the Royal Statistical Society, Series B, Methodological.* Vol. 41. 1-15.

[36] Desmond, A. F. and Chapman, G. R. (1993) Modeling Task Completion Data with Inverse Gaussian Mixtures. *Applied Statistics.* Vol. 42. No. 4. 603-613.

[37] Doganaksoy, Necip, Faltin, Frederick W., and Tucker, William T. (1991) Identification of Out of Control Quality Characteristics in a Multivariate Manufacturing Environment. *Communications in Statistics B: Theory and Methods.* Vol. 20. No. 9. 2774-2790.

[38] Doob, J. L. (1953) *Stochastic Processes.* New York: Wiley.

[39] Duncan, A. J. (1986). *Quality Control and Industrial Statistics.* 5 ed. Chicago: Irwin.

[40] Dupuy, Trevor N. (1987) Can We Rely Upon Computer Combat Simulations? *Armed Forces Journal International.* August. 58-63

[41] Edgeman, Rick. (1996) SPRT and CUSUM Results for Inverse Gaussian Processes, unpublished manuscript.

[42] Fellner, William H. (1990) AS 258 Average run length for cumulative sum scheme. *Applied Statistics* Vol. 39. 402–412.

[43] Ferguson, T.S. (1967) Mathematical Statistics. New York: Academic.

[44] Flury, B., and Riedwyl, H. (1988) *Multivariate Statistics A Practical Approach.* London: Chapman and Hall.

[45] Galpin, Jacqueline S., and Hawkins, Douglas M. (1984) The use of recursive residuals in checking model fit in linear regression. *The American Statistician.* Vol. 38. 94-105.

[46] Gan, F. F. (1989) Combined cumulative sum and Shewhart variance charts. *Journal of Statistical Computation and Simulation.* Vol. 32. 149-163.

[47] Gan, F. F. (1991) An optimal design of CUSUM quality control charts. *Journal of Quality Technology.* Vol. 23. 279-286.

[48] Gan, F. F. (1992a) Exact run length distributions for one-sided exponential CUSUM schemes. *Statistica Sinica.* Vol. 2. 197-312.

[49] Gan, F. F. (1992b) CUSUM control charts under linear drift. *The Statistician* Vol. 41. 71-84.

[50] Gan, F. F. (1993a), An optimal design of CUSUM control charts for binomial counts. *Journal of Applied Statistics.* Vol. 20. 445-460.

[51] Gan, F. F. (1993b) The run length distribution of a cumulative sum control chart. *Journal of Quality Technology.* Vol. 25. 205-215.

[52] Gan, F. F., and Choi, K. P. (1994) Computing average run lengths for exponential CUSUM schemes. *Journal of Quality Technology.* Vol. 26. 134-143.

[53] Goel, A. L. (1982) Cumulative sum control charts. *Encyclopedia of Statistical Sciences.* Vol. 2. 233- 241.

[54] Goel, A. L., and Wu, S. M. (1971). Determination of A.R.L. and a contour nomogram for CUSUM charts to control normal mean. *Technometrics.* Vol. 13. 221–230.

[55] Gold, Morris S. (1989) The geometric approximation to the CUSUM run length distribution. *Biometrika.* Vol. 76. 725-733.

[56] Goldsmith, P. L., and Whitfield,H. (1961) Average run lengths in cumulative chart quality control schemes (Corr: V3 p442). *Technometrics.* Vol. 3. 11-20.

[57] Goodman, Alan L. (1982) Cumulative sum control and continuous processes. *Proceedings of the ASQC Technical Conference.* 270- 274.

[58] Gordon, Louis, and Pollak, Moshe. (1994) An efficient sequential nonparametric scheme for detecting a change of distribution. *The Annals of Statistics.* Vol. 22. 763-804.

[59] Hawkins, Douglas M. (1981) A CUSUM for a scale parameter. *Journal of Quality Technology.*Vol.13. 228–231.

[60] Hawkins, Douglas M. (1987) Self-starting cusums for location and scale. *The Statistician* Vol. 36. 299-315.

[61] Hawkins, Douglas M.(1991) Multivariate quality control using regression adjusted variables. *Technometrics.* Vol. 33. 61-75.

[62] Hawkins, Douglas M. (1991) Diagnostics for use with regression recursive residuals. *Technometrics.* Vol. 33. 221–234.

[63] Hawkins, Douglas M. (1992) Evaluation of the average run length of cumulative sum charts for an arbitrary data distribution. *Communications in Statistics, series B* Vol. 21. 1001–1020.

[64] Hawkins, Douglas M. (1993) Robustification of cumulative sum charts by Winsorization, *Journal of Quality Technology.* Vol. 25. 248-261.

[65] Healy, J. D. (1987) A note on Multivariate CUSUM procedures. *Technometrics.* Vol. 29. 409-412.

[66] Hoel, Paul G., Port, Sidney C., and Stone, Charles J. (1972) *Introduction to Stochastic Processes.* Prospect Heights, Ill: Waveland.

[67] Hotelling, H. (1947) Multivariate quality control - illustrated by the air testing of sample bomb sights. *Techniques of Statistical Analysis* Ed. C. Eisenhart, M. W. Hastay and W. A. Wallis. New York: McGraw Hill.

[68] Jackson, J. Edward, and Mudholkar, G. S. (1979) Control procedures for residuals associated with principal component analysis. *Technometrics.* Vol. 21. 341-349.

[69] Jackson, J. Edward. (1985) Multivariate quality control. *Communications in Statistics* Vol. 14. 2657-2688.

[70] Jackson, J. Edward. (1991) *A User's Guide to Principal Components.* New York: John Wiley and Sons.

[71] Johnson, Norman L. (1961) A Simple Theoretical Approach to Cumulative Sum Control Charts. *Journal of the American Statistical Association.* Vol. 56. 835–840.

[72] Johnson, Norman L. (1963) Cumulative sum control charts for the folded normal distribution. *Technometrics.* Vol. 5. 451-458.

[73] Johnson, Norman L. (1966) Cumulative sum control charts and the Weibull distribution, *Technometrics*, Vol. 8. 481–491.

[74] Johnson, Norman L. and Leone, Fred C. (1962a) Cumulative Sum Control Charts: Mathematical principles applied to their construction and use. Part I. *Industrial Quality Control.* Vol. 18. No. 12. 15-21.

[75] Johnson, Norman L. and Leone, Fred C. (1962b) Cumulative Sum Control Charts: Mathematical principles applied to their construction and use. Part II. *Industrial Quality Control.* Vol. 19. No. 1. 29-36.

[76] Johnson, Norman L. and Leone, Fred C. (1962c) Cumulative Sum Control Charts: Mathematical principles applied to their construction and use. Part III. *Industrial Quality Control.* Vol. 19. No. 2. 22-28.

[77] Johnson, Norman L., and Kotz, Samuel. (1969) *Discrete Distributions.* Boston MA: Houghton Mifflin.

[78] Johnson, Richard A., and Bagshaw, Michael. (1974) The effect of serial correlation on the performance of CUSUM tests. *Technometrics.* Vol. 16. 103-112.

[79] Jorgensen, B. (1982) *Lecture Notes in Statistics: Statistical Properties of the Generalized Inverse Gaussian Distribution.* New York: Springer-Verlag.

[80] Jun, Chi-Hyuck and Choi, Moon Soo. (1993) Simulating the Average Run Length for CUSUM Schemes using Variance Reduction Techniques. *Communications in Statistics B: Simulation.* Vol. 22. No. 3. 877-887.

[81] Kemp, Kenneth W. (1967) A simple procedure for determining upper and lower limits for the average sample run length of a cumulative sum scheme. *Journal of the Royal Statistical Society, Series B, Methodological.* Vol. 29. 263-265.

[82] Khan, Rasul A. (1981) A note on Page's two-sided cumulative sum procedure. *Biometrika.* Vol. 68. 717-719.

[83] Lai, T. L. (1995) Sequential changepoint detection in quality control and dynamical systems. *Journal of the Royal Statistical Society, Series B, Methdological.* Vol. 57. 613–642.

[84] Lange, K. L., Little, R. J. A., and Taylor, J. G. (1989) Robust statistical modeling using the t distribution. *Journal of the American Statistical Association* . Vol. 84. 881-895.

[85] Lawless, J. R. (1982) *Statistical Models and Methods for Lifetime Data.* New York: Wiley.

[86] Lorden, G. (1971) Procedures for reacting to a change in distribution. *Annals of Mathematical Statistics.* Vol. 42. 1897-1908.

[87] Lowry, Cynthia A., Woodall, William H., Champ, Charles W., and Rigdon, Stephen E. (1992) A Multivariate exponentially weighted moving average control chart. *Technometrics.* Vol. 34. 46-53.

[88] Lucas, James M. (1976) The design and use of V-mask control schemes. *Journal of Quality Technology* Vol. 8. 1-11.

[89] Lucas, James M. (1982) Combined Shewhart-CUSUM quality control schemes, *Journal of Quality Technology* / Vol. 14. 51–59.

[90] Lucas, James M. (1985) Counted data CUSUMS. *Technometrics.* Vol. 28. 129-144.

[91] Lucas, James M. (1989) Control schemes for low count levels. *Journal of Quality Technology.* Vol. 21. 199-201.

[92] Lucas, James M., and Crosier, Ronald B. (1982a) Robust CUSUM. *Communications in Statistics A* Vol. 11. 2669–2687.

[93] Lucas, James M., and Crosier, Ronald B. (1982b) Fast Initial Response for CUSUM Quality Control Schemes: Give your CUSUM a Head Start. *Technometrics.* Vol. 24. No. 3. 199-205.

[94] Mandel, B. J. (1969) The regression control chart. *Journal of Quality Technology* Vol. 1. 1-9.

[95] McGilchrist, C. A., and Woodyer, K. D. (1975) Note on a distribution-free CUSUM technique. *Technometrics.* Vol. 17. 321-325.

[96] Montgomery, Douglas C. (1991) *Introduction to Statistical Process Control.* 2nd Ed. New York: Wiley.

[97] Montgomery, Douglas C., Keats, J. Bert, Runger, George C., and Messina, William S. (1994) Integrating statistical process control and engineering process control. *Journal of Quality Technology.* Vol. 26. 79–87.

[98] Montgomery, Douglas C., and Mastrangelo,Christina M. Some statistical process control methods for autocorrelated data *Journal of Quality Technology.* Vol. 23. 179-204.

[99] Moustakides, G. V. (1986) Optimal stopping times for detecting changes in distributions. *Annals of Statistics.* Vol. 14. 1379–1387.

[100] Munford, A. G. (1980) A control chart based on cumulative scores. *Applied Statistics.* Vol. 29. 252–258.

[101] Murphy, B. J. (1987) Selecting out-of-control variables with the T Multivariate quality control procedures. *The Statistician.* Vol. 36. 571-583.

[102] Nabar, S. P. and Bilgi, Shobar. (1994) Cumulative Sum Control Chart for the inverse Gaussian distribution. *Journal of the Indian Statistical Association.* Vol. 32. 9-14.

[103] Ncube, Matoteng M., and Amin, Raid W. (1990) Two parameter cuscore quality control procedures. *Communications in Statistics, Part A–Theory and Methods.* Vol. 19. 2191-2205.

[104] Ncube, Matoteng M., and Woodall, William H. (1984) A combined Shewhart-cumulative score quality control chart. *Applied Statistics.* Vol. 33. 259-265.

[105] Olwell, David H. (1996) *Topics in Statistical Process Control.* Ann Arbor: University Microfilms.

[106] Olwell, David H. (1997a) Statistical Process Control of Low Frequency Events. *Proceedings of the 5th Annual USMA-ARL Technical Conference.* West Point, NY: USMA.

[107] Olwell, David H. (1997b) Managing Misconduct: Statistical Process Control applied to sexual harassment. *1997 Proceedings of the Section on Quality and Productivity.* Alexandria, VA: American Statistical Association.

[108] Page, E. S. (1954) Continuous inspection schemes. *Biometrika* . Vol. 41. 100-115.

[109] Person, Ron (1996) *Special Edition, Using Windows 1995.* Indianapolis: Que.

[110] Pignatiello, Joseph J. Jr. and Runger, George C. (1990) Comparisons of Multivariate CUSUM Charts. *Journal of Quality Technology.* Vol. 22 No. 3. 173-186.

[111] Pollak, Moshe, and Siegmund, David. (1985) A diffusion process and its applications to detecting a change in the drift of Brownian motion. *Biometrika.* Vol. 72. 267-280.

[112] Quesenberry, Charles P. (1991) SPC Q charts for start-up processes and short or long runs, *Journal of Quality Technology* Vol. 23. 213–224.

[113] Quesenberry, Charles P. (1995) On properties of Poisson Q charts for Attributes. *Journal of Quality Technology.* Vol. 27. No. 4. 293-303.

[114] Regula, Gary A. (1976) Optimal Cumulative Sum Procedures to Detect a Change in Distribution for the Gamma Family. Ph.D. Dissertation, Case Western Reserve University.

[115] Reynolds, Marion R., Jr. (1975) Approximations to the average run length in cumulative sum control charts. *Technometrics.* Vol. 17. 65-71.

[116] Reynolds, Marion R., Jr. (1989) Optimal variable sampling interval control charts. *Sequential Analysis.* Vol. 8. 361-379.

[117] Reynolds, Marion R., Jr, Amin, Raid W., and Arnold, Jesse C. (1990) CUSUM charts with variable sampling intervals (C/R: p385-396). *Technometrics.* Vol. 32. 371-384.

[118] Ritov, Ya'acov. (1990) Decision Theoretic Optimality of the CUSUM Procedure. *Annals of Statistics* Vol. 18. No. 3. 1464-1469.

[119] Rocke, David M. (1989) Robust control charts. *Technometrics.* Vol. 31. 173–184.

[120] Ross, S. M. (1971) Quality Control under Markovian Deterioration. *Management Science.* Vol. 17.

[121] Ross, Sheldon M. (1990) Variance reduction in simulation via random hazards. *Probability in the Engineering and Informational Sciences.* Vol. 4. 299-309.

[122] Rowlands, R. J., Nix, A. B. J., Abdollahian, M. A., and Kemp, Kenneth W. (1982) Snub-nosed V-mask control schemes. *The Statistician.* Vol. 31. 133–142.

[123] Ryan, Thomas P. (1989) *Statistical Methods for Quality Improvement.* New York: Wiley.

[124] Srivastava, M. S. and Worsley, K. J. (1986) Likelihood ratio tests for a change in the multivariate normal mean. *Journal of the American Statistical Association* Vol. 81 199-204.

[125] Shewhart, Walter A. (1931) *Economic Control of Quality of Manufactured Product.* New York: Van Nostrand.

[126] Siegmund, David. (1985) *Sequential Analysis Tests and Confidence Intervals.* New York: Springer-Verlag.

[127] Siegmund, David. (1995) Using the generalized likelihood ratio statistic for sequential detection of a change-point. *Annals of Statistics.* Vol. 23. 255–271.

[128] Taylor, Howard M. (1968) The economic design of cumulative sum control charts. *Technometrics.* Vol. 10. 479-488.

[129] Tierney, Luke. (1990) *LISP-STAT: An Object Oriented Environment for Statistical Computing and Dynamic Graphics.* New York: Wiley.

[130] Tracy, N. D., Young, J. C., and Mason, F. L. (1992) Multivariate control charts for individual observations. *Journal of Quality Technology.* Vol. 24. 88-95.

[131] Vance, Lonnie. (1986). Average run lengths of cumulative sum control charts for controlling normal mean. *Journal of Quality Technology.* Vol. 18. 189–193.

[132] Van Dobben de Bruyn, C. S. (1968), *Cumulative Sum Tests: Theory and Practice.* London: Griffin.

[133] Vander Wiel, Scott A., Tucker, William T., Faltin, Frederick W., and Doganaksoy, Necip. (1992) Algorithmic statistical process control: Concepts and an application. *Technometrics.* Vol. 34. 286-297.

[134] Vardeman, Stephen, and Ray, Di-ou. (1985) Average run lengths for CUSUM schemes when observations are exponentially distributed, *Technometrics* Vol. 27. 145–150.

[135] Wade, M. R., and Woodall, W. M. (1992) A review of cause-selecting control charts. Unpublished manuscript.

[136] Wald, Abraham. (1944) On Cumulative Sums of Random Variables. *Annals of Mathematical Statistics*. Vol. 15.

[137] Wald, Abraham (1945) Sequential Tests of Statistical Hypotheses. *Annals of Mathematical Statistics*. Vol. 16.

[138] Wald, Abraham. (1947) *Sequential Analysis*. New York: Wiley.

[139] Wald, Abraham, and Wolfowitz, Jacob. (1948) Optimum character of the sequential probability ratio test. *Annals of Mathematical Statistics*. Vol. 19. 326-339.

[140] Waldmann, K. H. (1986) Bounds for the distribution of the run length of one-sided and two-sided CUSUM quality control schemes. *Technometrics*. Vol. 28. 61-67.

[141] Westgard, J. O., Groth, T., Aronson, T., and de Verdier, C. (1977) Combined Shewhart-CUSUM control chart for improved quality control in clinical chemistry. *Clinical Chemistry*. Vol. 23. No. 10. 1881–1887.

[142] Wetherill, G. Barrie. (1975) *Sequential Methods in Statistics*. London: Chapman and Hall.

[143] Woodall, William H. (1983) The distribution of run-length of one-sided CUSUM procedures for continuous random variables. *Technometrics* Vol. 25. 295-301.

[144] Woodall, William H. (1984) On the Markov chain approach to the two-sided CUSUM procedure. *Technometrics*. Vol. 26. 41-46.

[145] Woodall, William H. (1986) The Design of CUSUM Quality Control Charts. *Journal of Quality Technology* Vol. 18. No. 2. 99-102.

[146] Woodall, William H., and Ncube, Matoteng M. (1985) Multivariate CUSUM quality-control procedures. *Technometrics* Vol. 27. 285-292.

[147] Worsley, Keith J., (1979) On the likelihood ratio test for a shift in location of normal populations. *Journal of the American Statistical Association* Vol. 74. 365-367.

[148] Worsley, Keith J. (1986) Confidence regions and tests for a change-point in a sequence of exponential family random variables. *Biometrika* Vol. 73. 91-104.

[149] Xiao, H. (1992) A cumulative score control scheme. *Applied Statistics.*Vol. 41. 47-54.

[150] Yashchin, Emmanuel. (1985a) On a unified approach to the analysis of two-sided cumulative sum control schemes with head starts. *Advances in Applied Probability.* Vol. 17. 562–593.

[151] Yashchin, Emmanuel. (1985b) On the analysis and design of CUSUM-Shewhart control schemes. *IBM Journal of Research and Development.* Vol. 29. 377-391.

[152] Yashchin, Emmanuel. (1989) Weighted cumulative sum technique. *Technometrics.* Vol. 31. 321-338. Correction: Vol. 32. p. 469.

[153] Yashchin, Emmanuel. (1992) Analysis of CUSUM and other Markov-type control schemes by using empirical distributions. *Technometrics.* Vol. 34. 54–63.

[154] Yashchin, Emmanuel. (1993a) Statistical Control Schemes: Methods, Applications, and Generalizations. *International Statistical Review* Vol. 61. No. 1. 41-66.

[155] Yashchin, Emmanuel. (1993b) Performance of CUSUM control schemes for serially correlated observations. *Technometrics.* Vol. 35. 37-52.

[156] Yashchin, Emmanuel. (1995) Estimating the current mean of a process subject to abrupt change. *Technometrics.* Vol. 37. 311–323.

Index

Printed in the United Kingdom
by Lightning Source UK Ltd.
111035UKS00001B/102